Solar Selective Absorbers

Solar Selective Absorbers: Materials, Coatings, and Applications covers different harnessing technologies for solar energy, electromagnetic radiation including design principles, and different types of spectrally selective coatings.

The book emphasises experimental methods for synthesising these coating structures and characterisation techniques to quantify the absorber characteristics, suitable for solar thermal applications. It also provides a compact review of the literature on solar (spectrally) selective absorber materials and coatings.

- Summarises the materials, deposition and characterization techniques, and applications of solar selective absorbers
- Introduces solar selective absorber materials and photothermal technology
- Explores electrodeposition techniques and their use in the development of solar selective absorber films
- Provides an understanding of solar selective absorber materials and coating techniques in photothermal technology
- Explains sputtering techniques and their use in the development of ZrO_2/ZrC-ZrN/Zr tandem structure-based solar selective absorber films

The book is aimed at researchers and graduate students in solar energy, solar thermal conversion, and materials science.

Solar Selective Absorbers

Materials, Coatings, and Applications

Belal Usmani, Ambesh Dixit, Vivek Vijay,
and Rahul Chhibber

CRC Press
Taylor & Francis Group
Boca Raton London New York

CRC Press is an imprint of the
Taylor & Francis Group, an **informa** business

First edition published 2025
by CRC Press
2385 NW Executive Center Drive, Suite 320, Boca Raton FL 33431

and by CRC Press
4 Park Square, Milton Park, Abingdon, Oxon, OX14 4RN

CRC Press is an imprint of Taylor & Francis Group, LLC

ISBN: 978-1-032-64505-6 (hbk)
ISBN: 978-1-032-91577-7 (pbk)
ISBN: 978-1-003-56399-0 (ebk)

DOI: 10.1201/9781003563990

Typeset in Times LT Std
by Apex CoVantage, LLC

Contents

About the Authors

Belal Usmani received his bachelor and master degrees in physics from Shibli National College, VBS Purvanchal University, Jaunpur, Uttar Pradesh, India. He received a Master of Technology (Nanotechnology) degree from Aligarh Muslim University Aligarh, India, and his PhD degree from the Department of Physics, Indian Institute of Technology Jodhpur, Rajasthan, India. He joined Shibli National College, Azamgarh, affiliated with VBS Purvanchal University Jaunpur, in June 2018 as an assistant professor and currently works as an assistant professor in the Department of Physics at Shibli National College, MSD State University, Azamgarh. Prior to this, he worked as a project scientist in the Department of Materials Science and Engineering, Indian Institute of Technology Kanpur, India, from September 2017 to June 2018. His research interests are materials development and devices fabrication for energy applications and the theoretical calculation of electronic and optical properties of materials for energy applications.

Ambesh Dixit is a professor in the Department of Physics at the Indian Institute of Technology Jodhpur, India. He received his bachelor and master degrees in physics from the University of Allahabad, Allahabad, Uttar Pradesh, India. He previously worked at the Solid State Physics Laboratory (SSPL) as Scientist B, where he designed and developed a liquid phase epitaxy (LPE) system for the development of $Hg_{1-x}C_xdTe$ thin films. He also simulated the HgCdTe phase diagram for the understanding of its thermodynamic properties, essential to realising high-quality thin films for infrared (IR) detectors. He moved to Wayne State University (WSU), Detroit, Michigan, USA, and completed his PhD in multifunctional nitride and oxide materials. In 2009, he was the first to demonstrate to the scientific community across the globe that $FeVO_4$ is an intrinsic multiferroic, and he also designed an experimental set-up for controlling the magnetic ordering by applying an external electric field in contrast to the commonly used methodology of controlling ferroelectric ordering by applying an external magnetic field.

He joined IIT Jodhpur in July 2011 and created a research group on Multifunctional Materials from Theory to Experiment. The objective of the research is materials' design and engineering for various applications such as nanostructured materials for alternative energy (both solar photovoltaic and solar thermal) generation and storage, electronic materials for multistate memory devices, and multifunctional materials for strategic applications.

Vivek Vijay has been an assistant professor in the Department of Mathematics at the Indian Institute of Technology Jodhpur, Jodhpur, Rajasthan, India, since 2010. Prior to joining IIT Jodhpur, Dr Vivek spent a year and a half in the quantitative finance unit of Syntel Inc as Manager (Derivatives). He also worked at the Goa Institute of Management, Goa, India, as an assistant professor and with the Department of Mathematics at BITS Pilani, Goa Campus, as a lecturer. He received his PhD in statistics from the Department of Mathematics, Indian Institute of Technology

Bombay in 2007 and MSc (mathematics) from the Government College, Kota (presently Kota University), Rajasthan, India, in 2000. His research interests lie in the broad area of categorical data analysis, with some applications in collapsibility of contingency tables and forecasting of financial data, solar data, and health care data using machine learning and statistical techniques. Dr Vivek is a passionate writer on issues of social interest. Apart from his research work, he is also working on making mathematics simple by creating experiments in mathematics, especially for school students.

Rahul Chhibber received his Bachelor of Technology (BTech) and Master of Technology (MTech) degrees in mechanical engineering from Punjab Technical University, Jalandhar, Punjab, India, and the Indian Institute of Technology Roorkee, Uttarakhand, India, respectively. He received his PhD in mechanical engineering (welding) from the Indian Institute of Technology Roorkee, India. He has been an associate professor in the Department of Mathematics at the Indian Institute of Technology Jodhpur since 2019. Prior to joining IIT Jodhpur, he worked as an assistant professor in the Department of Mechanical Engineering, Thapar University, Patiala, Punjab, India. His research interests include fabrication and structural integrity assessment of dissimilar material joints, developing welding consumables for shielded metal arcs and submerged arc-welding processes. He has been involved with fabrication and hydrothermal investigations on fibre-reinforced polymer nano-composites.

Preface

The purpose of this monograph is to introduce the reader to study about solar selective absorber materials, coatings, and their application in photothermal technology. It is based on the PhD thesis which I completed in the Physics Department of the Indian Institute of Technology Jodhpur, Rajasthan, India. My aim is to introduce suitable text for senior undergraduate and graduate courses and for researchers and academics on the photothermal technology: solar selective absorber surfaces in physics and engineering.

The subject matter covered in this book is divided into seven chapters and two appendixes. The first chapter summarises the need of renewable energy and harnessing technologies of solar energy. The second chapter gives the general background of electromagnetic radiation, spectral selectivity for absorber surfaces, and a review of the previous work done so far on solar (spectrally) selective absorber materials and coatings. It highlights the challenges and difficulties for spectrally selective absorber materials and coating technologies. The subject matter of the present study is to overcome some of the challenges in the development of spectrally selective coatings. The third chapter deals with the experimental methods, including sample preparation and characterisation used in this work for spectrally selective absorber coatings. The synthesis and characterisation results for black chrome-graphite encapsulated FeCo nanoparticles composite (cremate) spectrally selective absorbers are summarised in the fourth chapter. Design, experimental fabrication, and characterisation results of zirconium carbonitride absorbers and zirconium metal reflector-based absorber–reflector tandem selective films are summarised in the fifth chapter. Chapter 6 summarises the solar selective absorber surfaces and their analysis methods for high-temperature ($T > 700$ °C), concentrating solar power tower technologies. Finally, Chapter 7 summarises and concludes the research work and provides input for future developments. The appendix discusses the experimental results on miscellaneous work. This summarises the experimental results on ZnO nanorod-based electrodes used for hydrogen evolution and storage studies.

Acknowledgements

I owe my deepest gratitude to my supervisors, *Ambesh Dixit, Vivek Vijay, and Rahul Chhibber*, who have provided me an opportunity to pursue PhD under their guidance at the Indian Institute of Technology Jodhpur (IITJ). Their immense support became the key success in my research work that I am presenting here. I extend my gratitude to Assistant Professor S. Harinipriya, presently at SRM University, Chennai, India, and Professor Rajiv Shekhar, presently at IIT Kanpur, who helped me in my research work. I am also thankful to my lab members, who have helped me throughout my research work at various levels.

I, hereby, thank all the scientists and professors who helped me, providing research guidance and support throughout my career. I thank Agilent Technology, Manesar centre, and the technical staff, for providing me an opportunity to carry out optical characterisations. I am grateful to Patra bhaiya's food and all the sips at Sangam Tea, which kept me consistent during work at the centre. I also thank staff members at IITJ from different laboratories – Mechanical Workshop, Chemistry, Electrical, and Microelectronics Laboratory – for support in assisting in the research efforts.

I pay my sincere gratitude to my parents, my sisters, brothers, uncle and his family, and aunt and her family for being with me all the time and silently supporting every step of mine, ignoring all my ignorance and mistakes. I was able to come so far because of their prayers, blessings, and support. I am indebted to Deepak, Suresh, Ritesh, and all my friends and colleagues for their moral support, suggestions, and appreciation that propelled me to travel this journey.

I sincerely express my deep sense of gratitude to my wife, Dr Asma Amjad Ghaznavi, and daughters Hamnah Usmani and Hadiyah Usmani for providing me with valuable time and support during the preparation of the manuscript.

I am indeed grateful to all the colleagues of the Department of Physics, Shibli National College Azamgarh, and friends of the college for their encouragement during the manuscript preparation.

I thank the Ministry of New and Renewable Energy (MNRE), Government of India, for financial assistance to carry out the experimental work during my PhD I thank the Department of Science and Technology (DST) and the Council of Scientific & Industrial Research (CSIR) for providing travel support for attending international conferences and presenting my work. Last but not least, I thank the Indian Institute of Technology Jodhpur and the Ministry of Human Resource Development (MHRD), Government of India, for providing me an avenue to shape my research career.

Symbols

Symbol	***Description***
Al	Aluminium
A	Actual Contact Area
Ar^+	Argon Ion
b_a	Anodic Tafel Constant
b_c	Cathodic Tafel Constant
Cu	Copper
E	Energy
E_{corr}	Corrosion Potential
e	Electrons
F	Faraday Constant
$\mathrm{G_{sc}}$	Solar Constant
H	Hardness
h	Plank's Constant
h_{max}	Maximum Displacement
I_{corr}	Corrosion Current
i_{corr}	Corrosion Current Density
K	Extinction Coefficient
k	Boltzmann Constant
m	Mass of the Electron
Ni	Nickel
n	Electron Density
n	Refractive Index
P_i	Protective Efficiency
P	Porosity
P_{max}	Maximum Load
R	Gas Constant
R	Reflectance
R_p	Polarisation Resistance
R_a	Average Surface Roughness
R_q	Root Mean Square Surface Roughness
r_0	Sun–Earth Distance
r_s	Mean Radius of Sun
S	Stiffness
T	Temperature
v	Poisson's Ratio
Y	Young's Modulus
α	Alpha
β	Beta
ε	Epsilon
π	Pi

λ	Wavelength
σ	Stefan–Boltzmann Constant
ε_0	Free Space Permittivity
η	Eta
δ	Sigma
%	Percentage
ω_p	Plasmon Frequency

Abbreviations

Abbreviation	*Full form*
AA	Air-Annealed
AC	After Corrosion
ACV	After Cyclic Voltammetry
AD	As Deposited
AFM	Atomic Force Microscopy
AM	Air Mass
ARC	Anti-Reflecting Coatings
ARTSSCs	Absorber Reflector Tandem Solar Selective Coatings
ASTM	American Society for Testing and Materials
ATA	After Tafel Analysis
BC	Black Chrome
CCD	Charged Coupled Devices
CSP	Concentrated Solar Power
CV	Cyclic Voltammetry
CVD	Chemical Vapour Deposition
DC	Direct Current
DNI	Direct Normal Radiance
DoF	Depth of Focus
DRA	Diffuse Reflectance Accessories
DSC	Differential Scanning Calorimetry
DTA	Differential Thermal Analysis
EDX	Energy-Dispersive X-ray
eV	Electron Volts
FeCo(C) NPs	Graphite Encapsulated Iron Cobalt Nanoparticles
FE-SEM	Field Emission Electron Microscopy
FTIR	Fourier Transform Infrared
FTO	Fluorine Doped Tine Oxide
FWHM	Full Width Half Maxima
HER	Hydrogen Evolution Reaction
hr	Hours
kW	Kilowatt
LSV	Linear Sweep Voltammetry
min	Minute
mm/y	Millimetre Per Year
mVS^{-1}	Millivolts Per Second
nm	Nano metre
ND	Neutral Density
NIR	Near Infrared
NRs	Nano Rods
PDF	Powder Diffraction File

PEC	Photo Electrochemical
PECVD	Plasma-Enhanced Chemical Vapour Deposition
PID	Proportional Integral Derivative
PLD	Pulsed Laser Deposition
PL	Photo Luminescence
PMT	Photomultiplier
PTFE	Polytetrafluoroethylene
PVD	Physical Vapour Deposition
RF	Radio Frequency
RMS	Root Mean Square
RT	Room Temperature
RVS	Raman Vibrational Spectroscopy
SCCM	Standard Cubic Centimetre Per Minute
SEM	Scanning Electron Microscopy
SPM	Scanning Probe Microscopy
SS	Stainless Steel
SSCs	Solar Selective Coatings
STA	Simultaneous Thermal Analyser
STP	Standard Temperature and Pressure
TG	Thermogravimetry
TW	Terawatt
UV	Ultraviolet
VA	Vacuum Annealed
WHO	Word Health Organization
XRD	X-ray Diffraction
YM	Young's Modulus
µm	Micro Metre

1 Introduction

The fast development and industrial growth across the globe, demand for energy stockpiles, and the continuous high demand for energy are causing the continuous depletion of natural resources such as oil, gas, and coal. On the other hand, the distribution, production, and use of fossil fuel energy have led to environmental degradation, causing global warming. The World Health Organization (WHO) has released data on the impact of direct and indirect effects of climate change on society, indicating the death of 1,60,000 people per year, which is supposed to be twice as that of 2020 [Mekhilef *et al.*, 2011]. Natural disasters such as flood, drought, and air pollution are taking place due to climate change and also affect the surrounding atmospheric temperature [Lior, 2008]. Conventional energy resources produce greenhouse gasses (CO_2, SO_2, CH_4, N_2O), which have contributed directly to global warming [Abdmouleh *et al.*, 2015]. The above-mentioned problems have forced the scientific community to contemplate alternative sources of energy. An alternative energy source should have low carbon emission and be more sustainable. Solar, wind, biomass, hydropower, and tidal energies are the alternative renewable energy sources, which can be explored to meet the ever-increasing energy demand [Schnitzer *et al.*, 2007].

Solar energy is one of the potential renewable energy sources. It has attracted attention because of its wide availability and environmentally benign nature [Liu *et al.*, 2013]. The rate of energy emitted by the sun is approximately 3.8×10^{23} kW, out of which approximately 1.8×10^{14} kW is intercepted by the earth, and the rest is reflected back to the atmosphere [Thirugnanasambandam *et al.*, 2010]. The intercepted energy is a thousand times our current power requirements. However, presently, 0.008 TW of solar energy is possible to use in different ways [Aman *et al.*, 2015]. Two major technologies are employed to convert the available solar energy into electricity and heat, as shown in Figure 1.1. One is photovoltaic (PV) technology, discovered by Edmund Becquerel (a French scientist) in 1893. In this, a semiconductor p–n junction device is used to convert solar energy into electricity directly [Solanki, 2013]. The second one is solar thermal technology, which converts solar energy into thermal energy, which can be used to generate electricity or to provide the thermal energy for other applications such as heating of water and space and cooling of buildings, heat requirements of industrial processes, and solar desalination [Mekhilef *et al.*, 2011]. Relatively cheaper and environment-friendly, the energy storage capability of solar thermal energy technology makes it more attractive as compared to the other renewable technologies, including photovoltaics, tidal, and wind, for large-scale energy applications. In solar thermal technology, heat can be stored during the sun hours, which can be used later on demand. This shows a higher dispatchability of electricity/power production using solar thermal as compared to the other possible renewable resources. Here, solar energy is converted into thermal energy using a solar collector/receiver. It is a major component of the solar thermal system, which

DOI: 10.1201/9781003563990-1

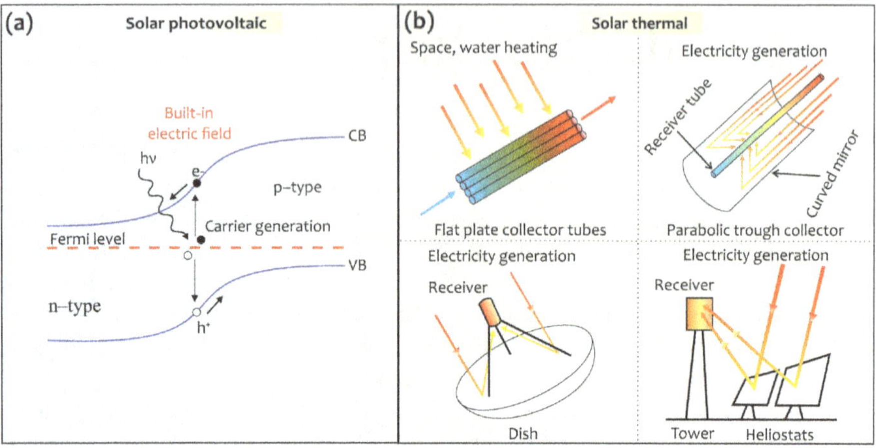

FIGURE 1.1 Two prominent techniques for solar energy conversion: (a) solar photovoltaic process, where incident energy is absorbed to generate charge carriers for direct electricity conversion and (b) different solar thermal including non-concentrating (e.g. space and water heating) and concentrating (parabolic trough, dish, and central tower receiver) technologies.

Note: Idea: www.volker-quaschning.de/

absorbs incident solar energy and converts it into heat. This heat is transferred into a fluid flowing through the collector for later processing [Selvakumar and Barshilia, 2012; Kalogirou, 2004]. Solar collectors can be classified as (i) non-concentrating or stationary and (ii) concentrating ones. In non-concentrating/stationary solar collectors, the intercepting area of solar radiation is the same as the absorbing area and is permanently fixed in position. These stationary collectors are further classified into (i) flat-plate collectors, (ii) stationary compound parabolic collectors, and (iii) evacuated tube collectors. In the case of a concentrating (sun-tracking) solar collector, solar energy is intercepted through a concave reflecting surface and focused onto a small receiver area, thereby increasing the radiation flux [Kalogirou, 2004]. The conversion efficiency of such collectors is limited because of conduction, convection, and radiation thermal losses from the absorber tube. These losses become significant at higher temperatures. Thus, efficient and economical utilisation requires the effective conversion of collected solar energy into the thermal energy, which will depend on the development of specific coatings, capable of converting the absorbed solar energy into thermal energy without any significant thermal loss. This poses materials' challenges to design spectrally selective absorbers, which may exhibit high absorbance (α) in the wavelength range of 0.3–2.5 μm (solar spectral range) and low thermal emittance (ε) in the wavelength range 2.5–25 μm (infrared region) at operating temperature ranges [Barshilia *et al.*, 2006a].

The spectrally selective absorbers can be categorised on the basis of their operating temperatures and related applications. These are (i) low-temperature absorbers ($T < 100°C$), (ii) mid-temperature absorbers ($100°C < T < 400°C$), and

(iii) high-temperature absorbers (T > 400°C) [Kennedy, 2002]. The mid-temperature absorbers are useful in low-temperature applications such as process heating for metal sintering and annealing, water desalination, and hot water. The high-temperature absorbers are used mainly in concentrating solar power (CSP) plants for generating electricity, and very few materials are available for such applications [Selvakumar and Barshilia, 2012].

The spectral selectivity concept was introduced by H. Tabor [Tabor, 1956, 1961], J.T. Gier, and R.V. Dunkle [Gier and Dunkle, 1958] in the mid-1950s, for the efficient solar thermal conversion of incident solar radiation. The initial work was focused on the low-cost electrodeposition process for absorber coating structures, including NiS–ZnS composite, nickel–nickel oxide (black nickel), chromium–chromium oxide (black chrome), and copper–copper oxide (black copper). These absorber coating structures were used in flat-plate collectors for water-heating applications. Solar absorptance of the electroplated black nickel on bright nickel-plated copper was $\alpha = 0.901$, and emittance was $\varepsilon = 0.05$ at T = 20°C [Tabor, 1956; Gier and Dunkle, 1958]. The extensive research work on solar selective absorber coatings was carried out in the late 1970s, during the oil crisis in 1973 and afterwards [Boffey, 1970]. Several reviews and research articles on commercial spectrally selective coatings are available, focusing on the development of these absorber structures and their possible applications [Agnihotri and Gupta, 1981; Seraphin, 1979; Kennedy, 2002; Bogaerts and Lampert, 1983a; Lampert, 1979a; Selvakumar and Barshilia, 2012; Cao *et al.*, 2014a; Amri *et al.*, 2014].

Various types of spectrally selective absorber coating structures have been investigated: (i) intrinsic or "mass absorbers", (ii) semiconductor–metal tandems and multilayer absorbers, (iii) metal–dielectric composite (cermet) absorbers, and (iv) surface texturing and selective solar-transmitting coating on a blackbody-like absorber. These spectrally selective coatings are produced using numerous techniques, such as wet chemical methods: electrodeposition, electrodeless deposition, anodisation, chemical conversion, thickness sensitive and insensitive spectrally selective paintings, sol-gel syntheses process, solution growth, spray pyrolysis, chemical vapour deposition (CVD), and physical vapour deposition (PVD) (evaporation, sputtering, pulsed laser deposition, ion plating) [Kennedy, 2002; Selvakumar and Barshilia, 2012]. Chemical methods are cost-effective, easily scalable, and, most importantly, low-temperature processes. However, these methods exhibit poor chemical and thermal stability, produce enormous waste materials, and, in some cases, methods are not environment-friendly [Konttinen *et al.*, 2003; Zhao and Wäckelgård, 2006].

A chemical vapour deposition process can be considered a probable choice for industrial or large-scale production, but the process depends on numerous parameters and imposes difficulties in maintaining the desired stoichiometries of constituent elements in the absorber structures [Kanu and Binions, 2010]. Physical vapour deposition (PVD), such as sputtering, has shown good reproducibility, low materials' consumption, and lesser environmental pollution as compared to CVD processes. However, PVD is less cost-effective because of its initial infrastructure-related high-cost requirements for large-scale production of absorber-coating structures and also the high operational cost and high-energy-density requirements for continuous production [Barshilia *et al.*, 2008a; Katumba *et al.*, 2008a].

The commonly used structures are metal–dielectric composite absorbers (cermet-based), multilayer absorbers, and absorber–reflector tandem structures. Various techniques and materials were used to develop a metal–dielectric composite (cermet) structures-based absorber coatings for mid-to-high temperature spectrally selective absorber applications [Cao *et al.*, 2014b]. Black chrome $(Cr\text{-}Cr_2O_3)$ is one of the most studied metal–dielectric composite materials, and electrodeposited black chrome is commercially developed by MTI on nickel-plated copper in the United States, by Chrome Coat on copper in Denmark, and by Energie Solaire on stainless steel in Switzerland [Brunold *et al.*, 2000]. TeKno Term Energi in Sweden and Showa in Japan have commercially developed nickel-pigmented alumina-based cermet selective coatings. Paint-based thickness-sensitive spectrally selective coatings were commercially developed by Solar etc-Z™ in Slovenia and by SolKoteHI/SORB-II™ in the United States. Various types of vacuum-deposited mid-temperature spectrally selective coatings are developed, but only a few of them are commercialised. For example, stainless steel-carbon (SS-C)-based coating has been developed and commercialised by Nitto Kohki in Japan; SS-AlN, CrN-Cr_2O_3, a-C:H/Cr, TiC/TiO_xN_y/AlN, and Ni-NiO spectrally selective coatings are developed and commercialised by TurboSun in China; by Alanod Solar, BlueTEC, Almeco-TiNOX, IKARUS Coatings in Germany; by PLASMA in Macedonia; and by S-Solar (Sunstrip) in Sweden. Mo-SiO_2, W/W-Al_2O_3/Al_2O_3, Mo-Al_2O_3, and W-Al_2O_3-based high-temperature spectrally selective coatings are developed by Angelantoni in Italy and by Siemens (formerly Solel) in Germany [Selvakumar and Barshilia, 2012].

So far, none of the existing previous selective absorber structures employed commercially demonstrated stability in air at or above 400°C. Therefore, developing a spectrally selective absorber coating structure, which is stable in the air beyond 400°C, is still a challenge. The high-temperature stability will require the development of materials for both individual and combined layers, which should exhibit an excellent adhesion between the substrate and adjacent layers and enhanced thermal and optical stabilities. In addition, the materials should also have high density and poor diffusion at higher temperatures to avoid the chemical degradation over the extended period at elevated temperatures. The thermal and mechanical resistances are also important for absorber materials to protect against the thermal cycling. The associated materials' challenges can be avoided by considering the high melting temperatures of refractory metallic and ceramic materials and their composites [Kennedy, 2010].

2 Radiation, Spectral Selectivity and Deposition Techniques of Surfaces, and Chronological Development of Spectral Selective Surfaces

Solar energy technology has gained attention in both academia and industry due to its abundance as a renewable energy source, which can be utilised for mitigating the everyday increasing energy demands. The aim of the solar energy technology is to harness the maximum for possible humankind applications. Solar thermal technology is the direct method to harness solar energy through solar collectors and receivers. The receiver consists of an absorber tube, for absorbing the maximum incident solar energy and minimising the thermal losses. The primary objective is to develop the thermally stable spectrally selective absorber coatings using low processing and maintenance cost. This chapter presents a brief review on solar electromagnetic radiation, optical properties such as absorptance and emittance, and the historical development of solar selective absorber materials for low and mid-to-high temperature applications. This will also include a section on commercially developed solar selective coatings and associated challenges/problems in the present context of solar thermal applications.

2.1 ELECTROMAGNETIC RADIATION

Electromagnetic radiation consists of electric and magnetic field vectors perpendicular to each other and also perpendicular to the direction of propagation. The energy will propagate along the direction of electromagnetic radiation and is known as the pointing vector. James Clerk Maxwell, a Scottish scientist, was the first to propose the theory of classical electromagnetism in 1862 and predict that light itself is a form of electromagnetic radiation. Electromagnetic radiation is divided into different energy bands based on the frequency distribution. The stipulated frequency range of electromagnetic radiation, such as solar radiation, which covers the ultraviolet–visible–near

DOI: 10.1201/9781003563990-2

infrared (UV-Vis-NIR) range (0.3–2.5 μm), and thermal radiation, from mid infrared (MIR) to far infrared (FIR) range (2.5–50 μm), are very important for solar thermal energy applications.

2.1.1 Thermal Radiation

Thermal radiation is the large-wavelength electromagnetic radiation, and a temperature-dependent radiation distribution is emitted by all bodies/objects. All objects, above absolute zero temperature, emit thermal radiation, whose wavelength and intensity depend on the temperature of the object and its optical properties. A blackbody is an ideal system that can absorb all the incident radiation, and to maintain a constant temperature, it emits an equal amount of energy simultaneously. The wavelength distribution of a blackbody at a given temperature, T K, is given by Plank's Law [Richtmyer and Kennard, 1947].

$$E_{\lambda b} = \frac{C_1}{\lambda^5[\exp(C_2 / \lambda T) - 1]} \qquad \text{Eq. (2.1)}$$

where $E_{\lambda b}$ is the energy per unit area per unit time per unit wavelength interval at wavelength λ. The subscript b indicates the blackbody, h is Plank's constant, and k is Boltzmann constant. $C_1 = 2\pi h C_0^2$ and $NO_2$are often called Plank's first and second radiation constants. The values of C_1 and C_2 are 3.7405×10^{-16} m^2W and 0.0143879 mK respectively. The total emitted energy is calculated by integrating Plank's radiation distribution over the desired wavelength range. The total energy emitted by a blackbody is found to be

$$E_b = \int_0^{\infty} E_{\lambda b} d\lambda = \sigma T^4 \qquad \text{Eq. (2.2)}$$

Here, σ is the Stefan--Boltzmann constant, equal to 5.6697×10^{-8} W/m^2K^4.

2.1.2 Solar Radiation

The sun is considered a spherical system with a diameter of 1.39×10^9 m and is approximately 1.5×10^{11} metres away from the earth. It emits about 3.84451×10^{23} kW energy using nuclear fusion reactions. The radiated solar energy spectrum consists of long-wavelength infrared radiation and very short-wavelength gamma radiation. However, most of this radiation is absorbed or scattered in the ionosphere, the ozone layer, or by atmospheric constituents such as nitrogen, oxygen, ozone, water vapour, and carbon dioxide. The intensity of solar radiation is constant outside the earth's atmosphere due to the sun and earth's spatial relation. The constant solar radiation is known as the solar constant, G_{sc}, which is $1367 \pm 1\%$ W/m^2 throughout the year. This solar constant is also defined as the energy received per unit area per unit time from the sun on a surface perpendicular to the direction of propagation of the incident radiation [Duffie and Beckman, 1991]. Solar radiation consists of about 8.03% UV, 46.41% visible, and the rest of 46.40% near infrared (NIR) fractions

(Figure 2.1 (b)). The sun's effective blackbody temperature can be calculated using the following equation [Iqbal, 1983].

$$T = \left[\frac{G_{sc}}{\sigma}\left(\frac{r_0}{r_s}\right)^2\right]^{1/4} \quad \text{Eq. (2.3)}$$

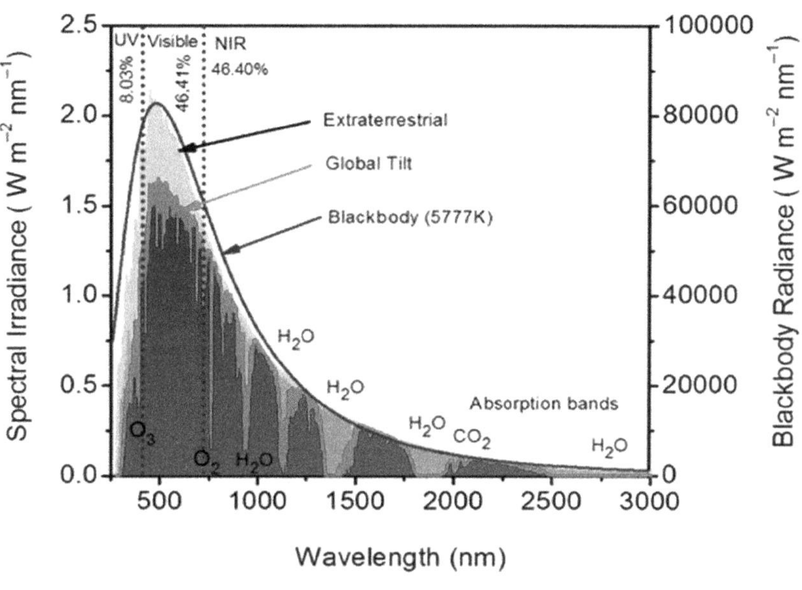

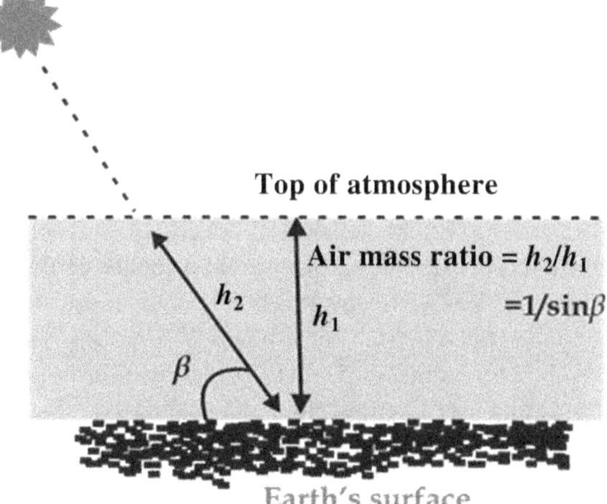

FIGURE 2.1 **a** spectral distribution of solar radiation – (i) extra-terrestrial radiation, (ii) total radiation incident on the inclined (37°) earth's surface, and (iii) blackbody radiation at sun's surface temperature 5,777 K with O_3, O_2, H_2O, and CO_2 absorption bands (source: RREDC, 2015). **b** the air mass ratio (AM) and its numeric analogue (source: Masters, 2004).

Where G_{sc} is solar constant and r_0 and r_s are the sun–earth distance and the mean radius of the sun, respectively. The effective blackbody temperature of the sun is 5,777 K. However, spectral distribution of the sun as a blackbody does not strictly follow the extraterrestrial solar spectrum; therefore, it cannot be used for characterising solar selective absorbers [Iqbal, 1983].

A fraction of solar radiation is absorbed by different atmospheric gases, giving an irregular and rough shape to the terrestrial spectrum, as shown schematically in Figure 2.1 (a). The absorption of extraterrestrial radiation is due to the resonance with vibrational energy levels of CO_2, NO_2, CO, O_2, and CH_4 gases. Simultaneously, the scattering takes place because of the gaseous molecules as well as atmospheric dust particulates [Sukhatme and Nayak, 2008]. The radiation is also characterised by air mass (AM), which is defined as the path length of the sun's rays passing through the atmosphere, divided by minimum sun rays' path length (Figure 2.1 (b)). Thus, an air mass 1 (AM1) implies that the sun is directly on the head. AM0 represents the same at no atmospheric conditions, that is in the extraterrestrial region. The AM1.5 is defined as an average solar spectrum at the earth's surface. The AM1.5 solar spectrum consists of 2% UV, 54% visible, and 44% infrared fraction in the solar spectrum [Masters, 2004] and is used for higher latitudes. For lower latitudes, a smaller AM solar spectrum is suggested. However, air mass variations have not shown much difference on the solar thermal performances for spectrally selective absorbers [Lind *et al.*, 1980].

There are two types of solar radiation: (i) beam or direct radiation termed Direct Normal Irradiance (DNI), directly received at the earth's surface without changing its direction, and (ii) diffuse radiation, received at the earth's surface from all parts of the atmosphere. The sum of these beams and diffuse radiations is called the global or total radiation [Sukhatme and Nayak, 2008]. Moreover, in concentrated solar thermal plants, electricity generation is directly affected by the availability and variability of solar resources and more specifically by DNI. Therefore, direct normal irradiation is crucial for concentrating direct sunlight to drive CSP plants [Blanc *et al.*, 2014; Nou *et al.*, 2016; Prasad *et al.*, 2015].

2.2 SPECTRAL SELECTIVITY OF ABSORBING SURFACES

The spectral distribution of the sun and blackbody spectrum at a given temperature needs to be understood in order to know the spectral selectivity of absorbing surfaces. After passing through the atmospheric layers, solar insolation on the earth's surface is confined in the 0.3 to 2.5 μm (UV-Vis-NIR) wavelength range, as shown in Figure 2.2 for clear atmospheric conditions. This spectrum corresponds to American Society for Testing and Materials (ASTM) G173 for air mass 1.5 (AM1.5). Figure 2.2 also shows blackbody spectra for 100°C, 200°C, and 300°C temperatures, calculated using Plank's law [Duffie and Beckman, 1991]. As the temperature rises, the total thermal radiation increases, and the peak wavelength shifts towards the shorter wavelength. The sun is a radiative source with a surface temperature of more than 5,500°C, emitting 95% of its output at wavelengths less than 2 μm, and a terrestrial receiver kept at 650°C (an upper temperature limit for the most technological applications) emits 95% of its thermal radiation at wavelengths more than 2 μm. Therefore, the spectral distribution of solar radiation and respective thermal losses are well separated on the wavelength scale, with approximately 5% overlap on either side of 2 μm, as illustrated in Figure 2.2 [Seraphin, 1979].

A selective absorber surface absorbs the maximum fraction of the incident solar energy while at the same time suppressing the thermal emittance in the infrared region. Neither perfect solar absorption nor perfect thermal suppression is possible [Seraphin, 1979]. For an ideal selective absorbing surface, the absorption is equal to one ($\alpha = 1$) and the emittance is equal to zero ($\varepsilon = 0$) at a given temperature, as shown in Figure 2.3. However, a real selective absorber surface has absorptance less than one ($\alpha < 1$) and thermal emittance greater than zero ($\varepsilon > 0$).

Solar absorptance and thermal emittance are two essential parameters to evaluate the spectral selectivity of an absorber surface. The solar absorptance is defined as the fraction of incident solar radiation absorbed by the absorber surface. According to Kirchhoff's Law, for a material which does not allow any transmittance to the incidence radiation, the spectral absorptance can be expressed in terms of total reflectance, $R(\lambda, \theta)$ [Duffie and Beckman, 1991],

$$\alpha(\lambda, \theta) = 1 - R(\lambda, \theta), \qquad \text{Eq. (2.4)}$$

where λ is the wavelength and θ is the incidence light angle from the normal of the absorber surface. The solar absorptance is calculated by weighting the spectral absorptance with spectral solar irradiance. The integration of weighted solar insolation *Isol* over the desired wavelength range is expressed as

$$\alpha(\theta) = \frac{\int_{\lambda_1}^{\lambda_2} [1 - R(\lambda, \theta)] I_{sol}(\lambda) d\lambda}{\int_{\lambda_1}^{\lambda_2} I_{sol}(\lambda) d\lambda}, \qquad \text{Eq. (2.5)}$$

where λ_1 and λ_2 represent the lower and upper solar wavelengths, respectively. Solar radiation reflectance $R(\lambda, \theta)$ is commonly measured in the range of 0.3–2.5 μm wavelength at near normal incidence using UV-Vis-NIR spectrophotometer.

Another parameter, thermal emittance, is used to characterise the performance of the solar selective absorber surface in the range of 2.5–25 μm wavelengths. The thermal emittance is calculated as a fraction of emitted radiation from the heated body as compared to the blackbody radiation at the same temperature. According to Kirchhoff's Law, for an opaque selective surface, the thermal emittance relates to the absorptance:

$$\varepsilon(\lambda, T) = 1 - R(\lambda, T). \qquad \text{Eq. (2.6)}$$

At a given temperature *T*, the total emittance is calculated as:

$$\varepsilon(T) = \frac{\int_{\lambda_1}^{\lambda_2} [1 - R(\lambda, T)] E_{\lambda b} d\lambda}{\int_{\lambda_1}^{\lambda_2} E_{\lambda b} d\lambda}, \qquad \text{Eq. (2.7)}$$

where $E\lambda b$ is the blackbody spectrum given in Equation (2.1). The denominator of Equation (2.7) is the total thermal energy radiated by a blackbody at temperature *T*.

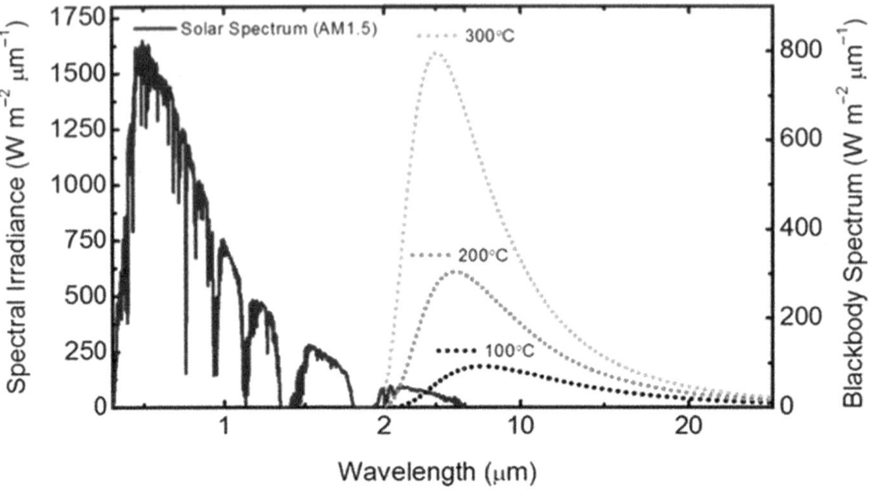

FIGURE 2.2 Solar spectral irradiance for AM1.5 in 0.3 to 2.5 μm wavelength range (source: RREDC, 2015). A blackbody radiation spectrum calculated using Plank's law for different temperatures (source: Duffie and Beckman, 1991).

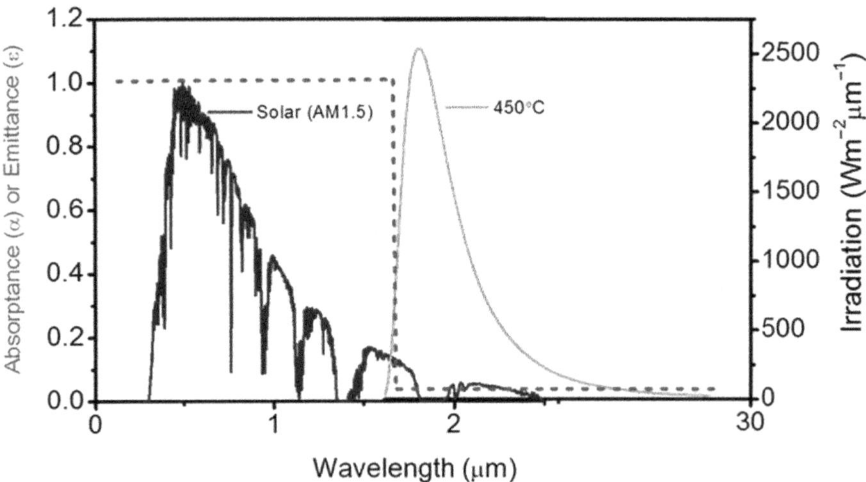

FIGURE 2.3 AM1.5 solar spectral irradiance in the wavelength range 0.3 to 2.5 μm (source: RREDC, 2015). The blackbody radiation for 450°C temperature (source: Duffie and Beckman, 1991). The red dashed curve represents the ideal spectral response for an absorber.

An efficient selective absorber is defined as having a high absorptance α in the solar spectrum range and low emittance ε in extended infrared range to reduce the thermal radiation loses, for temperature below 500°C. Ninety-eight per cent of this radiation occurs above 2-μm wavelength. The design of spectrally selective coatings is possible because of non-overlapping wavelength regions of these two properties: solar absorptance and emittance [Pettit and Sowell, 1976]. A quantity that has been used to characterise a spectrally selective coating is the ratio of the solar absorptance (α) to the thermal emittance (ε), known as the figure of merit or spectral selectivity of solar selective coatings [Jaworske, 2002]. A good spectrally selective coating might be expected to have the value of α/ε approximately 10, with large solar absorptance values [Pettit and Sowell, 1976]. For efficient spectrally selective structures, solar absorptance, α, should be as large as possible. Sandia research group has demonstrated that an increase in α is more effective in improving the operating efficiency than a corresponding decrease ε for focused solar collector system [Peterson and Ramsey, 1975]. For example, an increase of 0.05 in α (e.g. from 0.90 to 0.95) achieves the same increase in efficiency as it does by decreasing the emittance by 0.20 (e.g. from 0.30 to 0.10) [Pettit and Sowell, 1976].

2.3 DEPOSITION TECHNIQUES USED IN THE DEVELOPMENT OF COATINGS

Several deposition techniques are employed to develop the low-to-mid and high-temperature stable spectrally selective absorber coatings such as physical vapour deposition (PVD); chemical vapour deposition (CVD); plasma-enhanced chemical vapour deposition (PECVD); electrochemical methods; and sol-gel derived process such as dip-coating, spin-coatings, spray, and painting. In this section, the commonly used deposition techniques are discussed, which are especially employed for the development of spectrally selective structures.

2.3.1 Physical Vapour Deposition

Physical vapour deposition (PVD) techniques are based on the atomistic deposition process, in which the material is vaporised from a solid or liquid source in the form of atoms or molecules. The vaporised atoms are transported in the form of a vapour through a vacuum or low-pressure gaseous environment to the substrate, where it condenses. The PVD technique can be used for fabricating uniform films with variable thickness ranging from a few nanometres to micrometres [Mattox, 1997]. Various physical deposition techniques are explored by researchers, such as evaporation, sputtering, cathodic arc evaporation, and ion plating, for the development of spectrally selective absorber coatings.

2.3.1.1 Evaporation

Both thermal and electron beam evaporation techniques are employed to develop spectrally selective absorber coatings for photothermal applications. The raw material is heated in vacuum, where vaporised atoms and molecules are transported to the substrate. The rate of atom/molecule vapour transport can be controlled for getting the desired quality of the thin film structures. The necessary thermal heating is achieved by using a resistive coil or boat, made of refractory tungsten or molybdenum materials. In contrast to the resistive heating, used in case of thermal evaporation, a magnetically focused energetic electron beam is used in electron beam evaporation [Holland, 1956]. However, films deposited with evaporation do not bond well with the substrate and difficult-to-evaporate high-melting-temperature materials such as platinum and molybdenum. The major drawback of the evaporation techniques is the scaling issues for a large area deposition [Selvakumar and Barshilia, 2012]. There are limited reports on evaporation-based spectrally selective coatings. Niklasson *et al.* developed a $100^{\circ}C$ cermet coating with Al_2O_3 as an antireflection layer employing the co-evaporation of Co, and the absorptance and emittance values for optimised Co-Al_2O_3 cermet are approximately 0.94 and 0.04 at $\left(Cr\text{-}Cr_2O_3\right)$, respectively [Niklasson and Granqvist, 1983]. Several other cermet coatings such as Ag-Al_2O_3, Cr-Al_2O_3, and Au-Al_2O_3 have also been reported for solar thermal applications [Mckenzie, 1979].

2.3.1.2 Sputtering

Sputtering or sputter deposition relies on the sputtering or etching of target particles from their surface by the physical sputtering process. In this process, surface atoms are physically ejected from a solid surface when momentum is transferred from the atomic-sized energetic particles bombarding onto the target surface. It is an extensively employed technique for a large-area deposition for spectrally selective absorber coatings. In sputter deposition, control on film uniformity is relatively easier as compared to the evaporation deposition, especially in case of large-area depositions. Several coating structures have been investigated for absorber applications. For example, Cr-Cr_2O_3 cermet coating with Cr_2O_3 as an antireflection coating on nickel-coated stainless steel (SS) substrate has been developed using sputtering process, with enhanced solar thermal properties such as absorptance $\alpha = 0.98$ and emittance $\varepsilon = 0.08$ at 100°C [Fan and Spura, 1977]. A multilayer stainless steel–carbon (SS-C) coating is developed by Harding *et al.*, with a high absorptance value of approximately 0.94–0.95 and low emittance value of approximately 0.05–0.06 at room temperature [Harding, 1979]. Zhang reported an optimised Al-AlN-based three-layer cermet coatings, with absorptance $\alpha = 0.95$ and emittance $\varepsilon = 0.05$ at 80°C temperature [Zhang, 1999].

2.3.1.3 Arc Evaporation

A high current and low-voltage *dc* arc is used to evaporate the cathodic (cathodic arc) or anodic (anodic arc) electrode material. The evaporated electrode material is used to deposit on a substrate. The substrate is biased to accelerate the ionised materials' vapour. This process has extensively been used to deposit different metallic and compound thin films for various applications. However, one of the major disadvantages of this technique is the emission of macro particles from metal electrodes, which affects the optical properties of a solar selective coating. Yin *et al.* employed cathodic arc evaporation technique and prepared Al-AlN and a-C:H–SS (hydrogenated amorphous carbon–stainless steel) metal–ceramic structures on different substrates. Al-AlN metal–ceramic coating with AlN anti-reflecting coatings exhibited a good absorption of approximately 0.90 and an emittance of approximately 0.06 at room temperature. However, a-C:H-SS cermet with a-C:H antireflection coating exhibited a very low reflection in the visible region and thus high absorption of more than 0.95 in conjunction with relatively lower emittance values [Yin *et al.*, 1996].

2.3.1.4 Ion Plating

In this process, highly energetic atomic-sized particles are periodically bombarded on the substrate to deposit the desired thin film structures. The flux and mass of the bombarding species and the ratio of bombarding particles to the depositing particles on the substrate are important process variables for the deposition of thin film structures [Mattox, 1997]. Ion plating system has been employed to develop TiA/TiAlN/TiAlON/TiAlO spectrally selective coating by Lei *et al.*, coated on SS and Cu substrates and showed a high absorptance of 0.90 and a low emittance of 0.08–0.19 values. These structures are stable in air up to 650°C for 1 hour [Hao *et al.*, 2009].

2.3.1.5 Pulsed Laser Deposition

Pulsed laser deposition (PLD) is a very simple and elegant deposition technique and has been employed to deposit a broad range of materials such as optical coatings, high temperature superconducting, magneto-resistive, metallic and insulating thin film structures [Cheung, 1994; Tang *et al.*, 2010]. However, the major disadvantage of PLD technique is the available deposition area, which is relatively smaller than that of other deposition methods [Cheung, 1994]. Therefore, PLD technique has not been used extensively to deposit spectrally selective absorber coatings. Limited literature is available, where PLD technique has been employed to deposit cermet-based spectrally selective coatings. Chen *et al.* have studied the optical properties of Ti:Al_2O_3 selective coatings deposited using PLD and shown very low reflectance in the visible region and very high reflectance in the infrared region [Chen *et al.*, 2008].

2.3.2 Chemical Vapour Deposition (CVD) and Plasma-Enhanced CVD

In this process, atoms or molecules of interest are deposited by using the high-temperature decomposition of a chemical vapour precursor. The high-temperature thermal activation is used to achieve the desired decomposition of a chemical vapour precursor, whereas the reduction is carried out by using hydrogenous environment at elevated temperatures. However, in the case of plasma-enhanced CVD, plasma can be used to induce decomposition and reactions [Mattox, 1997]. In general, CVD technique has been used to produce coating structures for the electronic applications. However, different CVD processes have been investigated to synthesise the absorber coatings, including borides, carbides, oxides, and nitrides of transition and other metals [Lampert, 1979b; Kennedy, 2010]. For example, SiC:H thin films are grown from Ar/tetramethylsilane plasma-enhanced chemical CVD as absorber layers for high-temperature solar receiver applications, including the seven-layer stacks with alternating metal and SiC:H layers. The structure showed a high absorptance value of approximately 0.92 and a low thermal emittance of approximately 0.08 at 500°C [Glaude *et al.*, 2013].

2.3.3 Electrochemical Deposition Technique

Electrochemical deposition techniques such as electrodeposition (electroplating) and anodic oxidation have been extensively employed to develop spectrally selective absorber coatings. However, very few researchers have used electroless deposition technique for spectrally selective coatings. These techniques are briefly described for their applications in the development of spectrally selective coatings later in the chapter.

2.3.3.1 Electrodeposition Technique

Electrodeposition is an electrolytic process, where metal ions are moved in the solution by applying an electric field to deposit on one of the electrodes. The substrate to be coated is termed as a cathode in the electrodeposition cell. Electrodeposition has been extensively used to develop the absorber coatings, especially black chrome $(Cr\text{-}Cr_2O_3)$. In electrodeposition process, the performance of deposited solar selective coatings is affected by the several deposition parameters such as plating current density, bath temperature, plating time, and substrate. However, the environmental concern is a major challenge with electrodeposition. The process is well suited for electrically conducting substrates. However, insulating substrates such as glass and ceramics cannot be coated using electrodeposition [Atkinson *et al.*, 2015]. It has also issues of non-uniform film thickness and high thermal emittance.

2.3.3.2 Anodic Oxidation Technique

It is also an electrochemical process for the deposition of oxide films on the conducting substrates. In this process, electrons are repelled from an electrode, which oxidises the anode in an electrolyte to produce a coating. For example, the anodisation of aluminium is carried out in an aqueous solution of H_2PO_4 acid. Aluminium is used as an anode and immersed in the electrolyte and connected to the electrical circuit. A porous oxide layer is formed on the aluminium substrate. The anodised alumina is partially covered by metallic nanoparticles using electrodeposition to enhance the selectivity and absorptivity in the desired wavelength range.

Blain *et al.* reported the detailed experimental and theoretical study on solar selective surfaces, produced by anodisation and cobalt "Co" colouration of Al. The selective surfaces have shown a very high selectivity, with the absorptance α being 0.94 and emittance ε with a value of 0.1 [Blain *et al.*, 1985]. Andersson *et al.* have discussed a novel selective surface for an efficient photothermal conversion of solar energy, comprising Ni-pigmented anodic Al_2O_3 coatings on Al sheet [Andersson *et al.*, 1980].

2.3.4 Sol-gel Technique

Sol-gel technique is a chemical process and has been extensively used for the development of spectrally selective absorber coatings. Sol-gels are colloidal solutions of constituent elements, where the viscosity of the solution is controlled by using different solvents, required for the deposition of the desired film structures. Mostly, metal alkoxides are used for synthesising the gels. This process is attractive for receiver tubes and absorbers as it can be extended on intricate shapes and sizes, employing simple dipping or spray coatings techniques. The thickness of coatings can be tailored from few hundreds of nm to several micrometres. In terms of fabrication, there are different deposition options such as dip-coating, spin coating, spray coating, and flow- and roll-coatings, which can be used to coat a surface. However, industrially viable processes are spray-, flow-, and roll-coating techniques. Sol-gel processing does not usually need an expensive vacuum equipment and allows coatings of oxides at relatively low temperatures [Atkinson *et al.*, 2015]. Kaluza *et al.* prepared a novel black-coloured coating with CuCoMnOx compositions using sol-gel technique on Al substrate showing α value of 0.9 and ε value of 0.05 [Kaluza *et al.*, 2001].

2.3.5 Painting

Painting is an extremely simple process, where materials in liquidous states can simply be painted and dried to get the desired solar thermal properties. A paint solution usually consists of desired materials in organic solvents. The distribution of particles and their sizes are important to control the rheological properties of the paints, which may affect the solar thermal properties of the painted structures. In addition to the constituent elements, binder is also used in paint solutions to enhance the adhesion properties of paint materials with the substrates. The physical properties of painted film structures depend on the paint composition, solvent and binder components, their appropriate ratios, etc. These individual components are very important for the synthesis of thin-film structures with desired thermophysical properties. High-quality black paint, for example Pyromark, has been developed and still in use for high-concentration solar tower applications [Atkinson *et al.*, 2015]. $FeMnCuO_4$/silicon binder and PbS/polyacrylin resin are typical materials, used for painted coatings. The painting technique is an easy and low-cost process, suitable for large-scale applications. However, the major disadvantage of painted coatings is the relatively large emittance values and very poor bonding of selective surface with the substrate, causing peel off with time.

2.4 CHRONOLOGICAL DEVELOPMENT OF SPECTRALLY SELECTIVE ABSORBER SURFACES

The basic concept of spectrally selective coatings was developed by H. Tabor, J.T. Gier, and R. V. Dunkel for an efficient photothermal conversion during mid-1950s [Tabor, 1956, 1961; Gier and Dunkel, 1958]. Initially, coatings were successfully developed for flat-plate collectors, including black nickel, black chrome, and black copper oxides, using different fabrication methods. Electrochemical deposition has shown promise for such coatings with high absorptance and relatively higher emittance, as compared to the other physical processes. Values of solar absorptance (α) and thermal emittance (ε) of electroplated black nickel onto a bright nickel plated copper are 0.901 and 0.05 at T=20°C, respectively [Tabor, 1956; Gier and Dunkle, 1958]. The extensive research on solar selective absorber coatings was carried out in late 1970s, after realising oil crunch [Boffey, 1970]. The alternative renewable energy sources were attracting attention, and among them, solar thermal energy has been realised as the potential candidate. Thus, the development of selective surfaces became essential for the solar thermal energy applications. Several types of absorber surfaces such as intrinsic absorber, absorber–reflector tandem and semiconductor–reflector tandem, multilayer interference stacks, optical trapping systems, composite material films, and quantum size effect-based structures are developed by researchers across the globe. There is no such material, which may exhibit the ideal solar selective properties. However, there are several materials, which show very high selectivity. Due to the physical properties of materials, intrinsic solar selective properties can be found in two types of materials such as transition metals and semiconductors. Mostly, metals have a plasma reflection edge at about 0.3 μm, and it can be shifted towards infrared side by creating internal scattering centres. Moreover, tungsten (W) metal is one of the most wavelength-selective metals [Agnihotri and Gupta, 1981]. On the other hand, by using the highly degenerate semiconductors, it may be possible to shift plasma frequency in the infrared region [Dixit *et al.*, 2009]. Absorber–reflector tandem surface shows a high absorptance in the solar wavelength range and deposited on a highly reflective metal substrate. The onset of high absorption in the exterior coatings may be either of intrinsic nature or geometrically enhanced or the combined effect. Seraphin and Meinel have discussed the basic requirements of absorber–reflector tandem structures in detail [Seraphin, 1979]. In the case of semiconductor–reflector tandem selective surfaces, a thick semiconductor film is coated on an opaque metal having a high thermal infrared reflectivity. Semiconductor films with an energy band gap of about 0.5 eV (2.5 μm) to 1.26 eV (1.0 μm) could absorb radiation but are transparent to IR radiation. The spectral selectivity in multilayer stack is a result of multiple passes through the dielectric layers of the stack sandwiched between semi-transparent and totally reflecting surface. Meinel and Meinel have discussed the basic physics of multilayer stacks [Meinel and Meinel, 1977]. In optical trapping systems, surface texturing is a usual technique to get physical wavelength discrimination optical trapping of solar energy. Composite spectrally selective coatings, such as cermets, where small metal particles are embedded in a dielectric matrix, have good optical properties for spectrally selective coatings. Cermet surfaces absorb maximum solar spectrum due to the interband transition in metal and the small particle resonance, while they are transparent in the thermal infrared region. Resonant scattering depends upon both the size and optical properties of the particles and surrounding medium. Quantum size effects occur in ultrathin films of degenerate semiconductors, resulting into a high solar absorptance and simultaneously high thermal reflectance. The detailed mechanism of optical properties of spectrally selective coatings is discussed by Agnihotri and Gupta, Seraphin, and Kennedy [Agnihotri and Gupta, 1981; Seraphin, 1979; Kennedy, 2002]. In this section, we give an overview of historical developments of spectrally selective absorber surfaces – historical development of spectrally selective coatings categorised on the basis of temperature ranges such as low-, mid-, and high-temperature applications. We classified the temperature ranges based on their applications: low-temperature range is less than 200°C (Ni), mid-temperature range is more than 200°C but less than 350°C ($200°C < T < 350°C$), and high temperature range is greater than 350°C ($T > 350°C$).

2.4.1 Low-Temperature (T < 200° C) Spectrally Selective Absorber Materials

Normally, low-temperature spectrally selective surfaces are used in water-heating applications. Black nickel (Ni–Zn–S alloy) solar selective surfaces are used in absorber–reflector tandem structures since 1955 and commercialised long back for solar water-heating applications [Agnihotri and Gupta, 1981]. Chronological development of low-temperature spectrally selective surfaces is summarised in Table 2.1. The black nickel absorber structures are stable at relatively low temperatures. However, they start degrading in the humid and high-temperature environments, especially at or above 200°C and, thus, are poor humidity-resistant structures [Wäckelgård, 1998]. Black copper oxide (CuO) based spectrally selective absorber surfaces have also been developed in 1956 by Tabor *et al.* and later commercialised for industrial applications. High-temperature stability of the coatings is limited due to the defoliation and corrosion instability [Agnihotri and Gupta, 1981]. Table 2.1 illustrates the chronological developments of copper oxide spectrally selective coatings, using different synthesis techniques. Thickness-sensitive and thickness-independent spectrally selective surfaces, nanocomposite materials such as carbon-embedded dielectric matrices, and nickel-composite-based spectrally solar selective coatings are also developed for low-temperature photothermal applications. The development of such coatings is also summarised in Table 2.1.

TABLE 2.1
Historical developments of low-temperature solar selective coatings

Sr. No	Substrate	Deposition Technique	Absorptance (α)	Emittance (ε)	References
		Black Nickel (Ni–Zn–S alloy)			
1	SS	Electroplating	0.84	0.18	McDonald and Curtis, 1976
2	Nickel	Electroplating	0.96	0.007	Borzoni, 1976
3	316L SS	Electrodeposition	0.91	0.1	Lira-Cantú *et al.*, 2005
		Black Copper Oxide (CuO)			
4	SS	Electroplating	0.98	0.05	Bhowmik *et al.*, 2001
5	Copper	One-step chemical conversion method	0.94	0.08	Xiao *et al.*, 2011
6	Copper	Facile chemical oxidation	> 0.95	<0.07	Karthick Kumar *et al.*, 2013
		Thickness-Sensitive Spectrally Selective (TSSS) Coatings			
7	Aluminium	Coil-coating processes	0.92	0.38	Orel *et al.*, 1990

(Continued)

TABLE 2.1 *(Continued)*
Historical developments of low-temperature solar selective coatings

Sr. No	Substrate	Deposition Technique	Absorptance (α)	Emittance (ε)	References
8	Aluminium	Coil-coating processes	0.92	0.15	Orel *et al.*, 1991
9	Aluminium	Draw bar coater	0.90-0.92	0.2-0.25	Crnjak Orel *et al.*, 1996
10	Aluminium	Spray	0.90	0.20	Kunič *et al.*, 2011
		Thickness-Independent Selective Surfaces (TISS)			
11	Metal sheet	Sol-gel	0.90	0.3	Japelj *et al.*, 2008
		Carbon-Embedded Dielectric Matrix-Based Selective Surfaces			
12	Aluminium	Sol-gel	0.71 & 0.84	0.06 & 0.04	Katumba *et al.*, 2008b
13	Aluminium	Sol-gel	0.85	0.05	Roro *et al.*, 2012a
14	Aluminium	Sol-gel	0.84	0.2	Roro *et al.*, 2012b
		Nickel-Composite-Based Selective Surfaces			
15	Copper	Electrodeposition	0.94	0.08	Tharamani and Mayanna, 2007
16	FTO glass	Electrodeposition	0.83	0.14	Klochko *et al.*, 2015
		Novel Black-Coloured ($CuCoMnO_x$) Composite Selective Coatings			
17	Aluminium	Sol-gel	0.9	0.05	Kaluza *et al.*, 2001
		Nanogold-Containing Selective Coatings			
18	Copper	Solution chemical technique	0.86	0.09	Chang *et al.*, 2013

2.4.2 Mid-Temperature (200° C < T < 350° C) Spectrally Selective Absorber Materials

Mid-temperature spectrally selective absorber surfaces are employed for industrial applications such as process heat and desalination. Several spectrally selective absorber materials and structures are developed and commercialised for mid-temperature applications. Black chrome (BC) is one of the most studied spectrally selective absorber surfaces for photothermal applications. MTI (Materials Technology Inc.) has commercialised the BC solar selective coatings deposited on Ni-coated Cu substrate in the United States. Chrome Coat in Denmark and Energie Solaire in Switzerland have also developed BC-selective surfaces on Cu and SS substrates [Brunold *et al.*, 2000]. NASA Glenn Research Center has shown electroplated decorative finishing, known as BC, with desired solar selective properties. Here, BC was deposited on dull nickel-modified substrates or directly on rough surfaced substrates to increase the absorptance in the visible spectral region [McDonald, 1974]. The optimum plating time with current density 180 Ams/ft^2 (0.19 A/cm^2) is around one to two minutes, which resulted into the coating surfaces with optimal combination of the highest absorbance and the lowest radiative losses [McDonald and Curtis, 1975]. The best selectivity has been reported for 1.0-micron-thick BC-selective film on copper and 0.7-micron-thick BC on nickel substrates. In these plated BC structure, a suspension of metallic chromium particles has been observed with particulate size ranging between 0.05 and 0.30 microns. There is no degradation in reflectivity and microstructural properties in these coatings below 300°C temperature. However, BC-selective coatings degraded drastically after heat treatment in the range of $500°C - 600°C$ temperature even for one hour, both in air and vacuum [Lampert, 1978, 1979b]. The optical properties of BC selective films also depend on the microstructure of electrodeposited BC films, and a change in the microstructure of the absorber at elevated temperatures is the principal region for degradation of their solar thermal response [Ignatiev *et al.*, 1979a]. Degradation of electrodeposited BC coatings can be understood in two ways: (i) at low temperatures (< 300°C), a reduction in deposited hydroxide to oxide with desorption of H_2O and H_2, causing an upward shift of infrared response and, thus, enhancement in the infrared emittance and (ii) the high-temperature oxidation of the fine chromium crystallites comprising the coatings, resulting into the optical degradation in the visible range and thus a decrease in solar absorption [Zajac *et al.*, 1980]. The thermally treated BC-selective coatings showed minor changes below 300°C in air even for more than 100 hours of continuous heat treatment. However, above 300°C in both air and vacuum, there is a pronounced degradation. The main reason of this degradation is the diffusion of environmental oxygen, causing a conversion of metallic chromium particles into Cr_2O_3, leading to poor spectral selectivity [Lampert, 1980]. Optical properties of BC coatings also change due to substrate surface topography in near-infrared reflectance. However, the visible reflectance is insensitive to these changes [Axelbaum and Brandt, 1987; Sebastian *et al.*, 1997]. The selective absorption in BC coatings is dominated by the finely distributed metallic chromium particles inside the chromium oxide matrix. Lee *et al.* attributed the observed optical degradation to the conversion of chromium into chromium oxide at the surface of the BC coatings [Lee *et al.*, 2000] as deposited BC-selective surfaces have trivalent chromium hydroxide and chromium in the form of chromates. However, previous studies also suggested the presence of hydroxides and metallic chromium in these BC structures. The major component, chromium hydroxide, is converted to Cr_2O_3 after annealing at 400°C by losing water vapours [Anandan *et al.*, 2002]. Bayati *et al.* designed an electrochemical bath using chromium trivalent ion instead of toxic hexavalent ions [Bayati *et al.*, 2005]. Thick BC layers (~ a micron) are successfully electroplated on stainless steel substrates to form trivalent chromium salt, where BC structure consists of Cr, Cr_2O_3, CrO_3, and $Cr(OH)_3$. These structures exhibit good absorptance properties with value of approximately 0.97 and moderate emittance approximately 0.1 [Hamid, 2009; Eugénio *et al.*, 2011]. The developments of mid-temperature spectrally selective coatings are summarised in Table 2.2.

Among these mid-temperature spectrally selective coatings, aluminium oxide (Al_2O_3) (alumina) is another extensively studied ceramic material. This can be converted into metal–ceramic (cermet) structures using different metal particles in the host matrix. Several metals, such as Ni, V, Cr, Co, Cu, NiO, Ag, Si, and W, have been investigated in conjunction with aluminium oxide (Al_2O_3) for cermet structures [Möller and Hönicke, 1998]. For example, nickel (Ni)-pigmented alumina (Al_2O_3)-based selective surfaces are broadly studied and commercialised by

TABLE 2.2
Historical developments of mid-temperature solar selective coatings

Sr. No.	Substrate	Deposition Technique	Absorptance (α)	Emittance (ε)	References
		Black chrome ($Cr–Cr_2O_3$)			
1	SS	RF sputtering	0.9	0.1	Fan and Spura, 1977
		Metal-doped Alumina ($M-Al_2O_3$) cermet-based solar selective coatings			
2	Al	Anodisation	0.93–0.96	0.10–0.20	Andersson *et al.*, 1980
3	Al	Anodisation	0.96	0.20	Li, 2000
4	Al	Anodisation	0.93	0.03	Boström *et al.*, 2004
5	Al	Anodisation	0.94	0.1	Blain *et al.*, 1985; Zhang and Mills, 1992a; Cuevas *et al.*, 2014
6	Al	Anodisation	0.85	0.1	Möller and Hönicke, 1998
		Metal-Doped Aluminium Nitride (M-AlN) Cermet and AlN Multilayer-Based Solar Selective Coatings			
7	Glass	DC magnetron sputtering	0.93–0.96	0.03–0.04	Zhang *et al.*, 1998a
8	Glass and copper	Plasma-enhanced chemical vapour deposition (PECVD)	0.95	0.05	Rebouta *et al.*, 2012
9	Silicon (Si)	RF/DC sputtering	0.94	0.04	Du *et al.*, 2011
10	Quartz	Reactive DC sputtering	0.82–0.94	0.05–0.27	Y. Wu *et al.*, 2013
11	SS	Reactive DC sputtering	0.938	0.099	Feng *et al.*, 2015a
		NbTiON/SiON Selective Absorber			
12	Copper (Cu)	Magnetron sputtering	0.95	0.07	Liu *et al.*, 2012
		CrNxOy/SiO_2 Selective Absorber			
13	Cu and Si	DC magnetron sputtering	0.947	0.050	L. Wu *et al.*, 2013
		CrMoN/CrON			
14	SS and Al	Sputtering	0.90–0.92	0.13–0.15	Selvakumar *et al.*, 2013
15	Al	Magnetron sputtering	0.78	–	Gaouyat *et al.*, 2013
		Ultrafine Chromium Particles as Selective Absorber			
16	KBr	Evaporation	0.97	0.19	Granqvist and Niklasson, 1977
		Carbon-Silica Nanocomposite Selective Absorber			
17	Glass	Sol-gel	0.94	0.15	Katzen *et al.*, 2005
		Cr–O (Cr–CrO) Cermet Selective Coating			
18	Al	Planar DC magnetron	0.92–0.95	0.09–0.10	Lee, 2006
		CrxOy/Cr/Cr_2O_3 Selective Absorber			
19	Cu	Pulsed sputtering	0.899–0.912	0.05–0.06	Barshilia *et al.*, 2008a

TABLE 2.2 (*Continued*)
Historical developments of mid-temperature solar selective coatings

Sr. No.	Substrate	Deposition Technique	Absorptance (α)	Emittance (ε)	References
20	Al	Electrochemical process	0.923	0.06	Ding *et al.*, 2010
		Cu–Co–Mn–Si–O Black Selective Surface			
21	SS	Sol-gel (dip-coating)	0.95	0.12	Joly *et al.*, 2013
		TiN/TiSiN/SiN Selective Absorber			
22	SS	DC reactive magnetron sputtering	0.95	0.04	Feng *et al.*, 2015b
		MgO/Zr/MgO Multilayered Selective Absorber			
23	SS	E-beam evaporation	0.92	0.09	Nuru *et al.*, 2015
		W–SiO_2 and Nb–TiO_2 Cermet Selective Absorber			
24	SS	RF/DC sputtering	0.91/0.93	0.08/0.09	Wäckelgård *et al.*, 2015

Tekno Term Energi in Sweden and by Showa in Japan for photothermal applications. Developments of Ni-pigmented aluminium oxide spectrally selective coatings are summarised in Table 2.2 (see Sr. Nos 2, 3, 4, 5, and 6). In nickel pigmentation, nickel metal nanoparticles are introduced electrochemically into the porous alumina structures. The optical properties of hybrid structures depend on the nickel content to get the desired optical properties. However, nickel insertion along the pores of anodised aluminium is not necessary for obtaining a good optical performance [Salmi *et al.*, 2000], because selectivity in Ni-Al_2O_3 coatings is due to the composition of Ni/alumina composite and interference effect [Galione *et al.*, 2010]. Pigmentation with Ni acetate result in better emittance values as compared to Ni sulphate. However, both nickel-precursors-based structures are relatively insensitive to the solar absorptance values for these structures [Suzer *et al.*, 1998]. Metal-doped aluminium nitride (M-AlN) cermet-based solar selective coatings have also been extensively studied for photothermal application and developed for both mid- and high-temperature applications. The progress in M-AlN cermet selective absorbers is summarised in Table 2.2 (see Sr. Nos 7 to 11). These spectrally selective coatings are commercialised by China for the mid-temperature solar thermal application.

Several coatings with tandem structures such as cermet, multilayered, etc., have also been explored for mid-temperature solar thermal applications, which are summarised in Table 2.2 (see Sr. Nos 12 to 24). In multilayer metal–dielectric graded index solar selective coatings, there is an improvement of solar absorptance due to the destructive interference effects within the coatings [Farooq and Hutchins, 2002]. The main durability load for a solar absorber is the operating condition and stagnation temperature [Köhl *et al.*, 2004]. Spectrally selective coatings of carbon nanoparticles embedded in SiO_2, ZnO, and NiO matrices are fabricated by using the sol-gel technique on aluminium substrates. The NiO-based absorber surface exhibited an optimal solar selective behaviour, followed by ZnO and then SiO_2-based coatings. The NiO-based absorbers also exceeded the solar selectivity of SOLKOTE commercial paint coatings on the aluminium substrate [Katumba *et al.*, 2008a]. The metal–dielectric film structures exhibit a good spectral selectivity with solar absorption efficiency higher than 0.95 in the 400–1,200 nm wavelength range [Zhou *et al.*, 2012] and can be used for mid-temperature solar thermal applications.

2.4.3 High-Temperature (T > 350° C) Spectrally Selective Absorber Materials

High-temperature spectrally selective absorber materials are extremely important for their potential use in concentrated solar power (CSP) system for power generation. The efficiency of CSP plants strongly depends on the optical properties and thermal stability of the spectrally selective absorbers. The spectrally selective absorber materials should have high absorptance (> 0.95) and low emittance (< 0.10), not only at room temperature but also at high temperatures. These structures should also show a good thermal stability above 400°C in air and vacuum. The selective absorber materials should exhibit a high oxidation resistance and chemical inertness over the operating temperature range. The low emittance at high temperature is very important as the thermal radiative losses from the absorber are proportional to the fourth power of temperature (*T4*) [Duffie and Beckman, 1991]. Several high-temperature spectrally selective absorber materials are developed with time and are summarised in Table 2.3 (see Sr. Nos 1 to 35). Optical properties of the high-temperature spectrally

TABLE 2.3
Historical developments of high-temperature solar selective coatings

Sr. No.	Substrate	Deposition Technique	Absorptance (α)	Emittance (ε)	References
		MgO/Au Cermet Selective Absorber			
1	SS	RF sputtering	0.9	0.1	Fan and Zavracky, 1976
		M–Al_2O_3 (M = Au, Ag, Cr, Cu) Cermet Selective Surfaces			
2	Glass	Co-evaporation	0.90	0.05	McKenzie, 1979
3	Quartz	Co – evaporation	0.98	–	Craighead *et al.*, 1981
4	Cu	Co-evaporation	0.95	0.1	Niklasson and Granqvist, 1982, 1983, 1984
		Pt–Al_2O_3 Cermet Selective Absorber			
5	SS	RF Co-Sputtering	0.92	0.14	Vien *et al.*, 1985
		Co–Al_2O_3 Isotropic Cermet Sub-layers Selective Absorbers			
6	Cu and glass	E-beam evaporation	0.90	0.024	Zhang and Mills, 1992a, 1992b
		$MoSiO_2$/Al_2O_3 and Pt/Al_2O_3 Multilayer Selective Absorber			
7	Si, quartz, and alloys	RF sputtering	0.95 and 0.92	0.08 to 0.2 and 0.14	Schön *et al.*, 1994

TABLE 2.3 (*Continued*)
Historical developments of high-temperature solar selective coatings

Sr. No.	Substrate	Deposition Technique	Absorptance (α)	Emittance (ε)	References
		Al_2O_3 Cermet Selective Absorber			
8	Glass	Co-evaporation	0.955	0.032	Zhang *et al.*, 1996
		$Cr–Cr_2O_3$ and $Mo–Al_2O_3$-Based Layered Cermet Selective Absorber			
9	Cu and glass	DC sputtering	0.88 to 0.94	0.15 to 0.04	Teixeira *et al.*, 2001
		$Zr–ZrO_2$ Cermet Selective Absorber			
10	Glass	DC reactive sputtering	0.96	0.05	Zhang *et al.*, 2003
		$Ti–Al_2O_3$ Cermet Selective Absorber			
11	Nickel	Ion beam sputter deposition	0.93	0.09	Shumway *et al.*, 2003
		$Mo–SiO_2$ Cermet Selective Absorber			
12	SS and Glass	Sputtering	0.94	0.13	Esposito *et al.*, 2009
		$HfO_x/Mo/HfO_2$ Multilayer Selective Absorber			
13	SS and Cu	Magnetron sputtering	0.902–0.917	0.15–0.17	Selvakumar *et al.*, 2010
		$Ag–Al_2O_3$ Nano-Cermet Spectrally Selective Absorber			
14	Cu, Si, and glass	Magnetron sputtering	0.93	0.04–0.05	Barshilia *et al.*, 2011
		$Pt–Al_2O_3$ Cermet Selective Absorber			
15	SS	RF sputtering	0.98	0.05	Nuru *et al.*, 2012a
		$Al_xO_y/Pt/Al_xO_y$ Multilayer Selective Absorber			
16	Glass, Si, and Cu	e-beam vacuum evaporator	~ 0.94 ± 0.01	~ 0.06 ± 0.01	Nuru *et al.*, 2012b
		$Mo–SiO_2$ Cermet Selective Absorber			
17	Glass	Sputtering	0.95	0.15	Zheng *et al.*, 2013
		$Al_xO_y/Pt/Al_xO_y$ Multilayer Selective Absorber			
18	Cu	E-beam vacuum evaporator	0.951	0.09	Nuru *et al.*, 2014
		$Al_xO_y–AlN_x–Al$ Selective Absorber			
19	Quartz glass	DC Magnetron Reactive Sputtering	0.94	0.07	Yue *et al.*, 2003
		$TiAlN/TiAlON/Si_3N_4$ Tandem Selective Absorber			
20	Cu	Magnetron sputtering	0.95	0.07	Barshilia *et al.*, 2006b
		$Ti_{0.5}Al_{0.5}N/Ti_{0.25}Al_{0.75}N/AlN$ Selective Absorber			
21	Si	RF/DC magnetron co-sputtering	0.945	0.04	Du *et al.*, 2011

(*Continued*)

TABLE 2.3 (*Continued*)
Historical developments of high-temperature solar selective coatings

Sr. No.	Substrate	Deposition Technique	Absorptance (α)	Emittance (ε)	References
		TiAlN/TiAlON/SiO_2 tandem Selective Absorber			
22	Cu and glass	DC magnetron sputtering	0.955	0.05	Rebouta *et al.*, 2012
		HfMoN(H)/HfMoN(L)/HfOH/Al_2O_3 Tandem Selective Absorber			
23	SS	Magnetron sputtering	0.94–0.95	0.13–0.14	Selvakumar *et al.*, 2012
		Ti/AlTiN/AlTiON/AlTiO Multilayer Selective Absorber			
24	SS	Magnetron sputtering	0.933	0.16–0.17	Barshilia, 2014
		TiAlCrN/TiAlN/AlSiN Multilayer Selective Absorber			
25	SS	cathodic arc	0.91	0.07	Valleti *et al.*, 2014
		SS–(Fe_3O_4)/Mo/TiZrN/TiZrON/SiON			
26	SS	Reactive DC/RF magnetron sputtering	0.92	0.08	Liu *et al.*, 2014a
		Al_2O_3:W Cermet Layers and AlSiN/AlSiON Bilayer Selective Absorber			
27	SS	Magnetron sputtering	0.93–0.95	0.07–0.01	Rebouta *et al.*, 2015
		NbTiON/SiON Selective Absorber			
28	Cu	Magnetron sputtering	0.95	0.07	Liu *et al.*, 2012
		Cr–Al–O Tandem Nano-multilayer Composite Selective Absorber			
29	SS and Si	Cathodic arc ion plating	0.924	0.21	Liu *et al.*, 2014b
		AlCrNO Selective Absorber			
30	SS	Cathodic arc plating	0.93	0.20	Gong *et al.*, 2015
		CrAlO Selective Absorber			
31	SS	Cathodic arc ion plating	0.918–0.924	0.154–0.170	Liu *et al.*, 2015
		SS–AlN Cermet Selective Absorber			
32	Glass	Sputtering	0.94–0.95	0.04–0.05	Zhang, 1998; Zhang *et al.*, 1998b
		ZnO/Black ZnO Selective Absorber			
33	Glass	Sputtering	0.90	0.26	Brett *et al.*, 1986
		Mo–Si_3N_4 Selective Absorber			
34	SS and Si	Sputtering	0.926	0.017	Céspedes *et al.*, 2014
		Carbon Nanotube-Based Selective Absorber			
35	SS, Cu, Ni, and Cr	Spray	>0.92	<0.01	Abendroth *et al.*, 2015

selective absorber films usually start degrading due to the diffusion of environmental oxygen into absorber layers and oxidising of the metal content inside the absorber layers. In addition, the substrate material may start diffusing into the absorber structures at higher temperatures (>400ºC) [Cheng *et al.*, 2013].

Furthermore, an electrodeposited Co_3O_4 selective absorber coating has shown an absorptance of more than 0.90 and an emittance less than 0.2, with an enhanced thermal stability at 650°C for 1,000 hours [McDonal, 1980]. New Si–Ge selective coating structures based on multi-scaled semiconductor nanostructures are designed and fabricated, exhibiting a high absorptance for short-wavelength photons and an efficient reflection for long-wavelength photons [Moon *et al.*, 2014]. A black oxide structure with cobalt oxide nanoparticles, embedded in a dielectric matrix through a scalable spray coating, has shown a high thermal efficiency of 88.2%. These structures are thermally stable and do not show any degradation even at 700°C in the air after 1,000 hours of annealing [Moon *et al.*, 2015].

Large numbers of spectrally selective coatings have been developed for high-temperature solar thermal applications, as discussed before. However, few of them such as Mo-SO_2, Mo-Al_2O_3, and M–AlN, where M: SS, W, and Mo cermets, have been successfully commercialised and are being used in evacuated tubes for solar thermal power generation. The Mo-Al_2O_3 cermet coated receiver tubes are produced by Luz International Ltd., the United States, and are used in Solar Energy Generating System power plants. The coatings exhibited an absorptance of 0.96 and an emittance of 0.16 at 350°C. However, this Mo-Al_2O_3 cermet structure has a limited thermal stability in the air up to 300ºC [Lanxner and Elgat, 1990]. SS, W, and Mo-AlN-based selective coatings are commercially developed by TurboSun, exhibiting approximately 0.92–0.94 solar absorptance and approximately 0.08–0.10 emittance at 350°C. Mo-SiO_2 and W-Al_2O_3-based spectrally selective coatings are developed by Archimede Solar Energy (ASE), Italy, with a relatively enhanced solar absorptance of more than 0.94 and emittance less than 0.13 at 580°C.

3 Solar Selective Absorber Surfaces

Electrodeposition and Sputtering Techniques and Characterisation Techniques

There are various synthesis techniques, which can be used for fabricating spectrally selective structures, as reviewed in Chapter 2. This chapter will discuss in detail about the experimental techniques used for the synthesis of spectrally selective coatings, used in this thesis work. Section 3.1 includes the electrochemical deposition and DC/RF magnetron sputtering processes. These spectrally selective absorber coatings are further characterised to understand the structure and microstructure, constituent elements, thickness of coatings, optical properties such as absorptance and emittance and mechanical and corrosion properties. Further, thermal stabilities of these structures have also been investigated using thermal gravimetric analysis and cyclic heating. The details of characterisation techniques, used for this work, are described in Section 3.2.

3.1 SAMPLE PREPARATION TECHNIQUES

3.1.1 Electrochemical Deposition

The electrochemical deposition, also known as electrodeposition, is one of the metal and alloys' thin-film deposition techniques. In this technique, ions from electrolytic (aqueous, organic, and fused-salt) solutions are reduced and transferred in the presence of an electric field to form a film on the desired substrate. The electrodeposition cell consists of an anode, cathode, and electrolyte. Furthermore, anodes are divided into two categories: one is sacrificial and another is a permanent anode. The sacrificial anodes are to be deposited, and the permanent anode is employed to complete the electrical circuit and does not participate in the electrochemical process. An electrolyte is the ionic conductor, which allows ions to get transported from one electrode to another instead of free electrons to complete the internal electrical circuit. After applying bias/electrical potential, the cations and anions in the electrolyte will move towards the cathode and anode. Moreover, ions conduct through the electrolyte to draw the electrical current in the internal circuit, which is complemented by an equivalent electronic current in the external circuit. The metallic cations are attracted towards the cathode through the electrolyte, where

DOI: 10.1201/9781003563990-3

excess electrons on the cathode (workpiece) neutralise the cations into metallic ones. These reduced cations form a metallic film on the surface of this negatively charged workpiece (i.e. cathode) [Paunovic and Schlesinger, 2006; Lou and Huang, 2006]. The electrochemical setup is illustrated schematically in Figure 3.1(a). The system is kept on a hot plate to achieve the desired temperature during electrodeposition

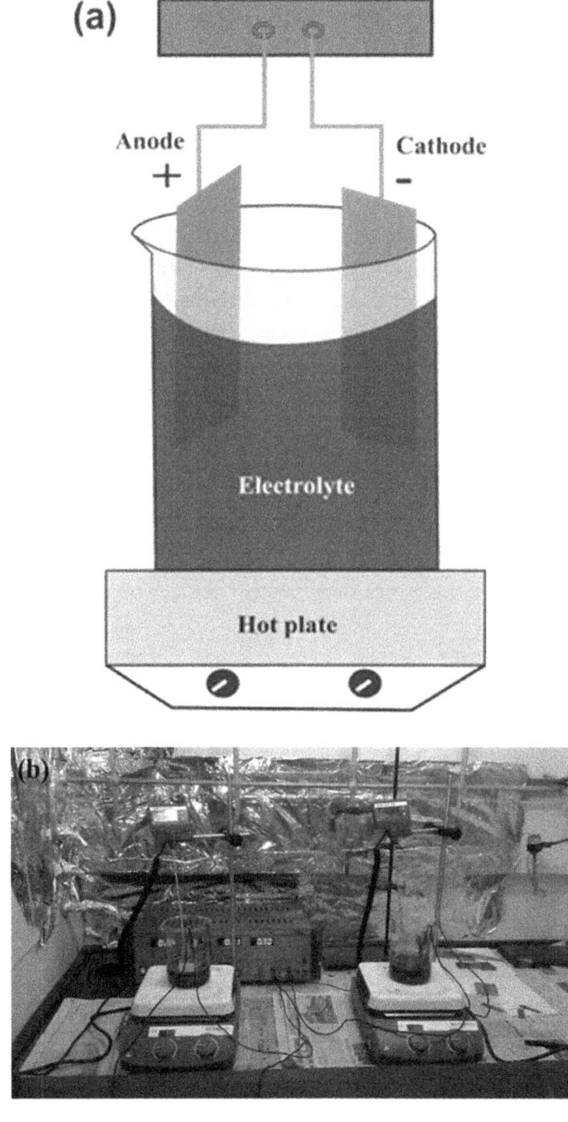

FIGURE 3.1 (a) Schematic diagram of an experimental setup for electrodeposition. (b) An electrochemical deposition experimental setup in our lab used to deposit spectrally selective absorber films.

process and connected with DC power supply to provide the desired voltage/current for potentiostatic/galvanostatic electrodeposition process. The actual experimental setup is shown in Figure 3.1(b), which is used to deposit spectrally selective absorber films on different substrates such as stainless steel, copper, and aluminium in the present work.

Redox (reduction/oxidation) reactions are electron transfer reactions in the electrolysis process. The electrode, gaining electrons, oxidises another electrode, which is losing electrons in the process. Cations are reduced at cathode, accepting electrons, which flow towards the anode (Equations (3.1) and (3.2)).

$$M^{n+} + ne^{-} \rightarrow M \tag{3.1}$$

$$M \rightarrow M^{n+} + ne^{-} \tag{3.2}$$

The electrodeposition of metal oxides is mainly performed in alkaline aqueous solutions containing metal complexes under oxidising and reducing conditions. In each case, the metal ions dissociate from metal complexes and precipitate on the electrode as oxides, and the process can be explained in two steps. The first step is the metal dissolution, which is an electrochemical process (Eq. 3.2). The second step is the chemical process between metal and hydroxide ions, happening on the surface of the substrate.

$$M^{n+} + OH^{n-} \rightarrow M(OH)_{n} \tag{3.3}$$

The electrodeposited oxides may consist of complex compositions such as metal alloys, metal oxides, and metal hydroxides. Further, a controlled heat treatment can be used to convert such composites into having the desired composition for probable applications.

3.1.2 RF/DC Sputtering Technique

Sputtering is a physical deposition process, where materials (neutral atoms or molecules) are etched from a solid surface, known as the target (due to the transfer of momentum between an energetic gas particle (ions) and the target surface). Grove was the first one, who observed the sputtering process in a DC gas discharge tube in 1852, and now this process is frequently used to deposit the desired thin-film structures.

The important parameter for the sputtering process is the yield (ratio of the number of ejected atoms to the number of bombarding ions) of the sputtered materials. This is the function of chemical binding of the target material and collision energy impact. There are two types of sputtering sources: one is glow discharge source and another is an ion beam source. The following section discusses glow discharge sputtering and RF/DC magnetron sputtering processes. Different types of sputtering systems are developed for thin-film depositions. Among them, the first one is the DC diode sputtering system and others are the advancement of this DC diode sputtering system [McClanahan and Laegried, 1991].

3.1.2.1 DC Diode and Triode Sputtering

DC diode sputtering has a vacuum chamber with two electrodes, appropriate pumping, and the gas flow arrangement, as illustrated in Figure 3.2(a). The target material is fixed at the cathode, to be sputtered, and held at a negative potential. Substrates are placed on the anode. Positively charged Ar^+ ions are bombarded on the target, and ejected neutral target atoms/molecules are deposited onto the substrate [Shah, 1995]. An optimum gas density is required to maintain the glow discharge. In low gas density, electrons are impacted on the anode without ionising any argon atom; however, at a higher gas density, electrons do not have the sufficient energy to ionise the argon atom. The secondary electrons, which are produced by the accelerated ions, play a major role in ionising atoms in the discharge. The main disadvantages of DC diode sputtering are low deposition rates, high gas densities, high discharge voltages, lack of versatility, and suitability for conducting targets only [Rossnagel, 1991]. A heated filament is used to inject electrons directly into the plasma to further enhance the ionisation of plasma, as shown in Figure 3.2(b), which enhances the electron density and, thus, providing higher deposition rates. This is called the triode sputtering, because of the additional electron source filament arrangement. The major problems with triode sputtering are (i) its scaling up and (ii) gas species interaction with the filament, which may create impurities during deposition.

3.1.2.2 Radio Frequency (RF) Sputtering

RF sputtering is employed to produce insulating thin films and coatings such as dielectric oxide films [Davidse and Maisel, 1966; Davidse, 1967a, 1967b]. The main limitations of an RF sputtering system are the high cost power supply and the complicated matching network, which is utilised to lower the reflected power during the sputtering process. A 13.56-MHz radio frequency power supply is used with impedance matching network and blocking capacitor, as illustrated in Figure 3.3. Electrons are oscillating between two electrodes with the frequency of applied power [Rossnagel, 1991]. A net negative DC bias is applied to attract the ions for sputtering and the resulting deposition of target materials on a desired substrate [Est and Westwood, 1998]. In an RF sputtering, the capacitive coupled plasma is formed near the target, which makes no difference in the target surface whether the target is electrically conductive or insulating. However, RF sputtering rate of insulating materials is relatively poor due to the lower sputter yield of insulators. The power is divided between the two electrodes in an RF sputtering process, and the RF sputtering power at the cathode is half of the power delivered in DC sputtering. This low effective power at the cathode is the main reason for low deposition rates in RF sputtering processes as compared to that in an equivalent DC process. The major problem with DC diode, triode, and RF sputtering systems is the non-uniform plasma densities, causing non-uniform current density and film thickness during the deposition process.

3.1.2.3 Magnetron Sputtering

In a conventional sputtering process, the target is fixed with cathode plate and bombarded with energetic ions, produced in the available plasma. The target materials are etched out during bombardment process, and they settle down in the form of a film on the substrate. The secondary electrons, which are generated by the target at

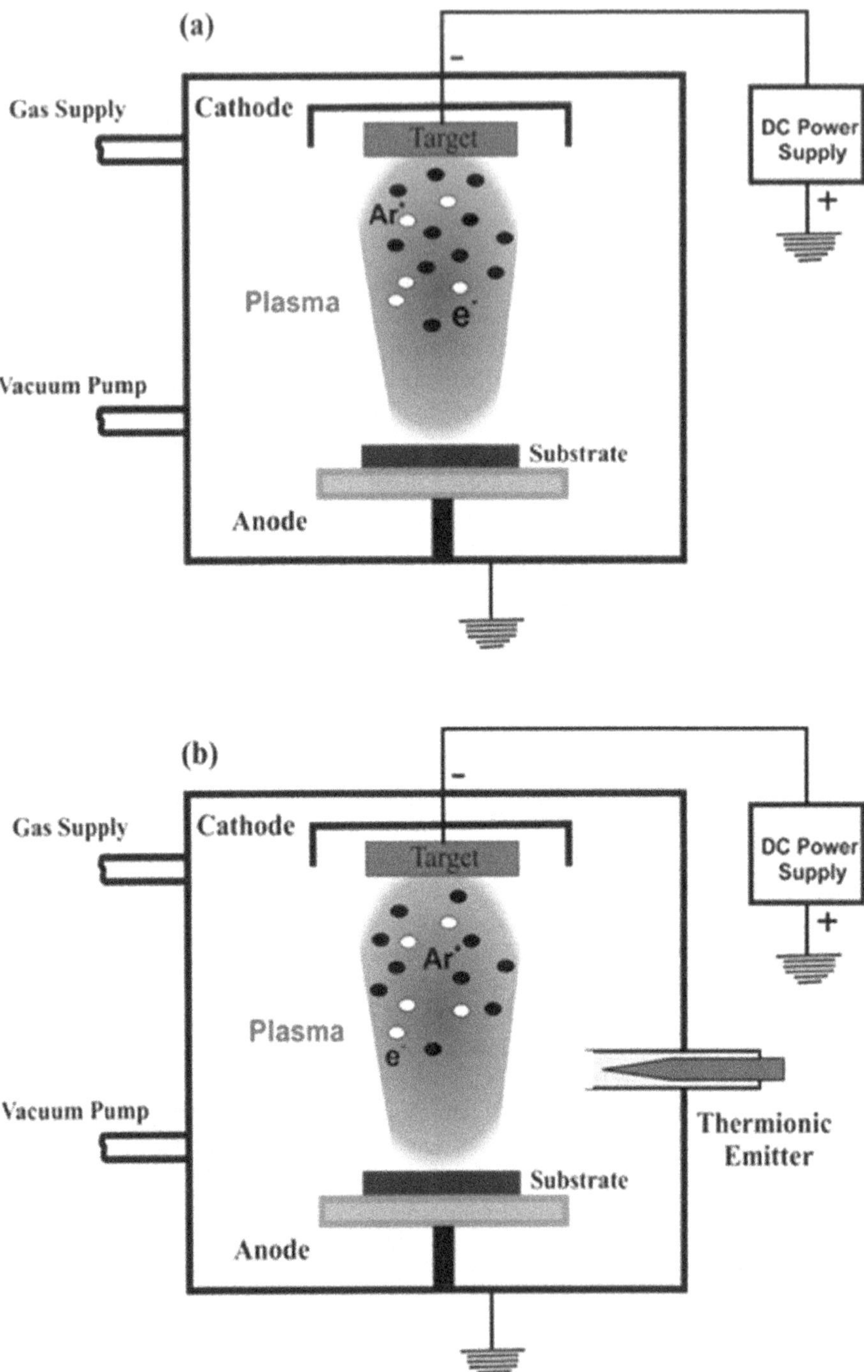

FIGURE 3.2 Schematic of (a) a DC diode sputtering system and (b) a DC triode sputtering system. Source: Waite *et al.* (2010).

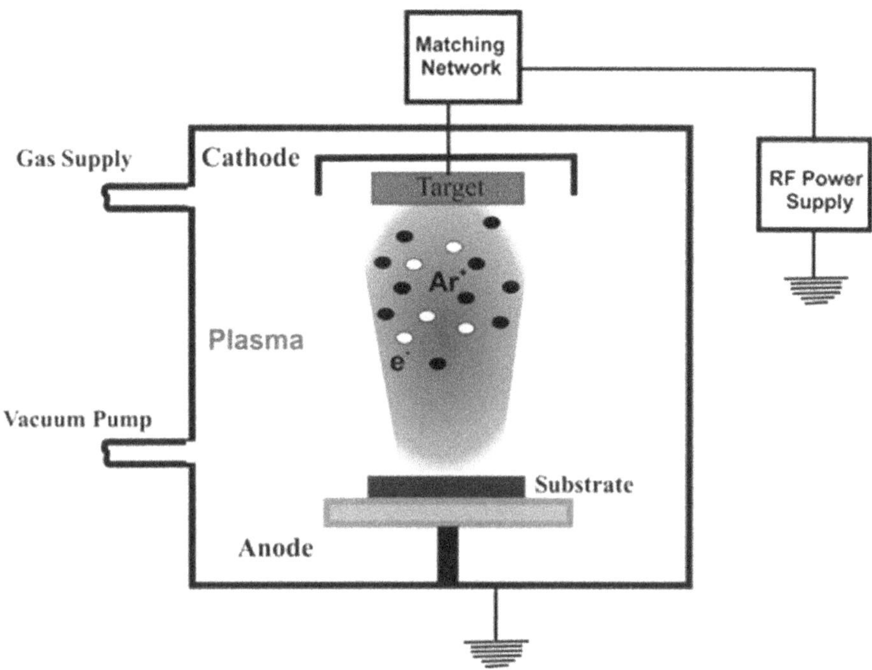

FIGURE 3.3 Schematic diagram of an RF sputtering system, with respective legends.

Source: Waite *et al.* (2010)

the time of bombardment, also play a crucial role in sustaining plasma during synthesis. However, conventional sputtering procedure has constraints like low deposition rate, low ionisation capacity, and more substrate-heating impacts. These constraints can be minimised by using the magnetron sputtering system in place of conventional DC and RF sputtering systems [Kelly and Arnell, 2000]. The confinement of electrons near the surface target can be increased with the help of magnetic field so that the plasma density increases in magnetron sputtering. The first demonstration of a magnetron sputtering was done by Dutch experimental physicist Frans Michel Penning in 1935. Further, Peter J. Clark had developed a sputtering source, known as sputtering gun, in 1968, and J.S. Chapin designed a planar magnetron in 1974 and patented the same [Waite *et al.*, 2010].

3.1.2.3.1 Planar Magnetrons

The arrangement of a planar magnetron sputtering system is identical to that of DC and RF sputtering system. However, it has a fixed magnetic arrangement at the back of the cathode to confine the electrons near and above the targets (Figure 3.4(a)) [Waits, 1978]. The orientation of the magnetic field is radial and overhead the cathode. Hence, the $\mathbf{E} \times \mathbf{B}$ drift forms a circular path along the target. This localised nonuniformity creates a circular region on the target surface, which is known as the racetrack, resulting in a poor utilisation of target $\left(20\% \text{ to } 45\%\right)$. In the 1980s,

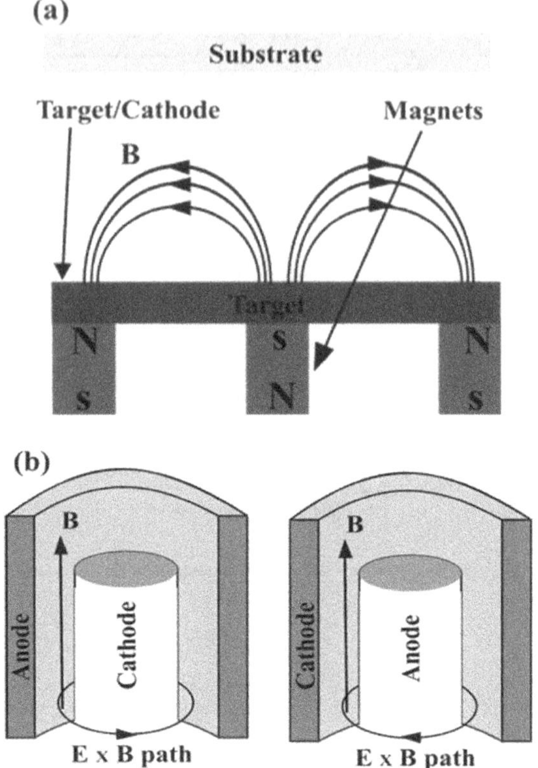

FIGURE 3.4 Schematic diagram of (a) planar magnetron sputtering system and (b) cylindrical magnetron sputtering system.

Source: Waite *et al.* (2010)

Michael Wright and Terry Beardow designed the rotatable cylindrical magnetron to solve the problem of low-target utilisation. Cylindrical magnetrons have been extensively employed in the coating of complex-shaped substrates [Siegfried *et al.*, 1996]. Figure 3.4(b) illustrates the cylindrical magnetrons, where the cathode is along the axis, and the anode is a concentric tube. The inverted arrangement of the same is known as a hollow magnetron.

3.1.2.3.2 Unbalanced Magnetrons

Keeping in mind the above-said problems in planar/conventional magnetron sputtering, Window and Savvides developed a new design, known as an unbalanced magnetron in which magnets of different strength are used. Employing the unbalanced magnetron sputtering system could increase the ion bombardment on the target, and thus growing films become denser and preferentially oriented [Pradhan and Shah, 2002]. There are two types of the unbalanced magnetron: type I has a strong central magnet and type II has strong outer magnets, as explained in Figure 3.5

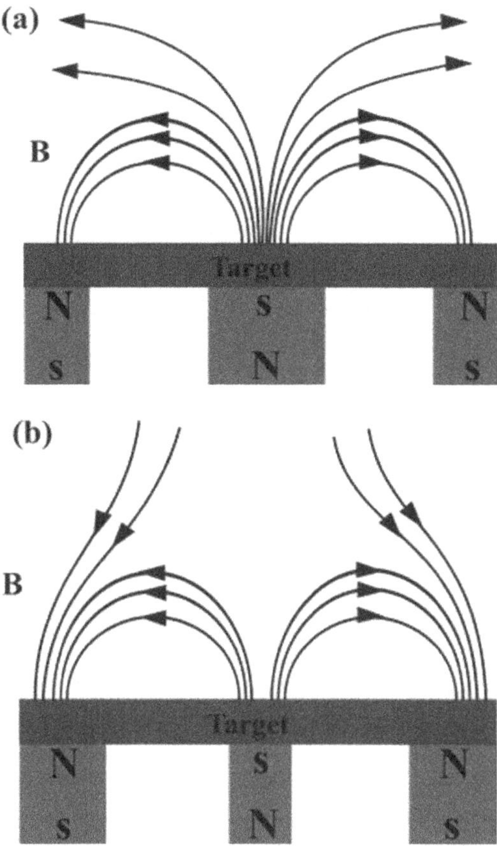

FIGURE 3.5 Schematic diagram of unbalanced magnetron sputtering system (a) type I and (b) type II.

Source: Waite *et al.* (2010)

[Window and Savvides, 1986]. One of the major disadvantages of the unbalanced magnetron is that the plasma is not uniform and thus may lead to the non-stoichiometry and non-uniform deposition in the final structures.

3.1.2.3.3 Reactive Sputtering

The reactive sputtering process has been extensively investigated since 1980 due to the possibility of getting an easy synthesis process for compound films, such as nitrides, oxides, carbides, or their combinations, using conventional metal targets [Musil *et al.*, 2005]. Reactive sputtering of the metallic target with DC power supply is less expensive than RF power supply because of matching network and low yield with RF sputtering process. In reactive sputtering, gasses are in a molecular state. Normally, reactive gasses have low atomic masses and do not contribute effectively in the sputtering of the target atoms. That is why argon is used as a sputtering gas in conjunction with a small fraction of reactive gas during the deposition process. The problems with reactive sputtering are the creation of a compound layer on the target itself and afterward target surface charging, which finally reduces the sputtering yield. The sputter rate decreases due to the conversion of target surfaces into respective insulating compounds, which is also known as poisoning of the target surface. The detailed specification of a combined DC/RF sputtering system, which is used for the present work, is described next.

The combined DC/RF sputtering system has a box-type stainless steel vacuum chamber (process chamber), with 550 mm (width) × 450 mm (height) × 550 mm (length) dimensions. The inner view of this vacuum chamber (process chamber) is shown in Figure 3.6. A thin stainless steel sheet is used inside the chamber to prevent the deposition on the chamber walls. It has four water-cooled 4" diameter magnetron sources, with an optimised inclination to include the 6" deposition region, as shown in Figure 3.6 (arrows are marked for easy identification). An electro-pneumatically operated source shutter has been used with each magnetron source to open or close the required sputtering source during the sputtering process. The substrate holder is made of non-magnetic stainless steel material, integrated with a rotational mechanism. A substrate heater is placed below the substrate holder, which can be used to heat the substrates up to 800°C, using a digital PID controller in conjunction with a K-type thermocouple to measure the temperature.

A 400 (W) × 400 (D) × 450 (H) mm^3 box-type 304 SS load lock chamber is integrated with the main deposition chamber. The inside view of load lock chamber is shown in Figure 3.7. A magnetic transfer mechanism with sample loading unit is used for transferring the substrate manually between the load lock and the process chamber. Linear positioning is controlled by sliding an external sleeve, which is magnetically coupled with transporters and sample holder inside the system.

Two RF power supplies (at 13.56 MHz frequency and 600 Watts maximum) (make M/S. SEREN, the United States) are installed with the sputtering system. This has

FIGURE 3.6 An inside view of the vacuum chamber (process chamber) of a combined RF/DC magnetron sputtering system in our lab used to deposit spectrally selective absorber films.

an auto matching network to feed RF power at resonance condition to the magnetron source during the thin-film deposition process. In addition, two, 2 kW DC power supplies are integrated with the system and can be used for DC sputter deposition. A 300 W, 13.56 MHz RF power supply (M/S, SEREN, US make) is available with the system for substrate biasing and pre-cleaning. This combined RF/DC sputtering system is configured with a direct drive rotary pump (model FD-60) having 1,000 lit/min free air displacement capacity. The pump is fitted with an anti-suck back device to protect the system from back streaming of rotary pump oil during power failures. Leybold turbo molecular pump with model TURBOVAC 1100C is configured with the system in line with a rotary pump. Figure 3.8 shows the front view of Hind High Vacuum (HHV) RF/DC magnetron sputtering system.

FIGURE 3.7 An inside view of load lock chamber of a combined RF/DC magnetron sputtering system.

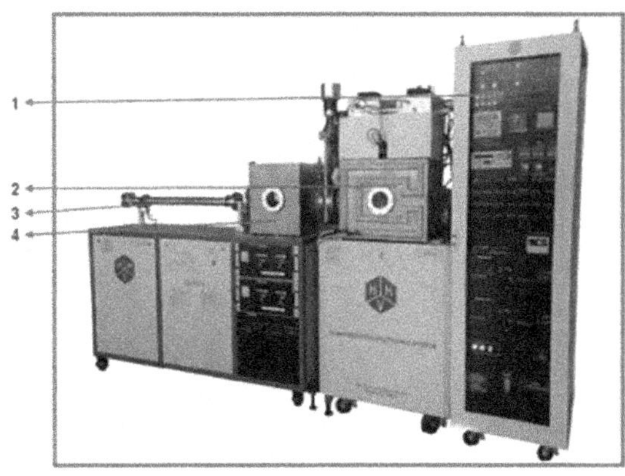

FIGURE 3.8 Front view of the combined RF/DC magnetron sputtering system.

3.2 HEAT TREATMENT

Heat treatment is a process, used for baking materials, at the desired heating rate and for a period of time. This can be used to change the properties of the materials as well. The objective of the heat treatment is to achieve the desired physical, chemical, and mechanical properties, etc. [Rajan *et al.*, 1992]. Thermal-analysis-related equipment, used to understand the thermal degradation of spectrally selective coatings in air and vacuum, is described later in Section 3.3.

In this work, heat treatment processes have been explored to understand the materials' response against the thermal treatment under both air and vacuum/inert conditions. For air treatment, a tabular furnace (Nabertherm GmbH), shown in Figure 3.9, has been used for annealing specially in air, wherever required for the thermal treatment of spectrally selective coatings. The temperature of this furnace can be varied up to 1,100°C at different heating rates.

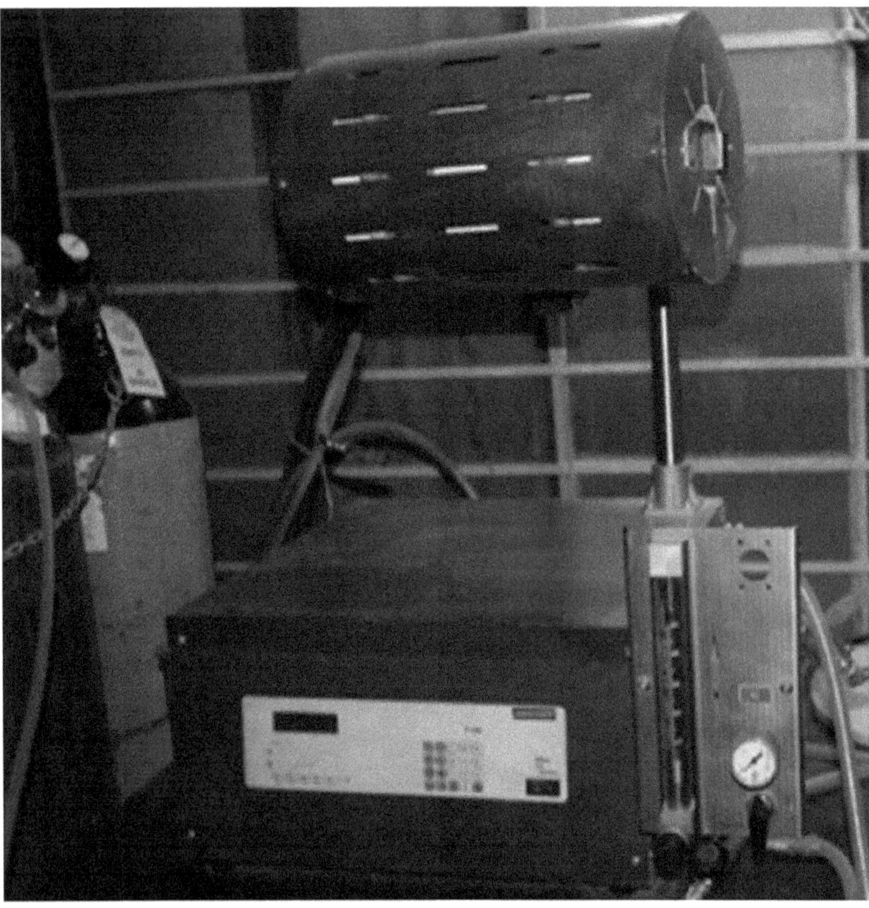

FIGURE 3.9 Tabular furnace (Nabertherm GmbH) in our lab used for annealing of spectrally selective absorber films in this study.

Vacuum annealing of the spectrally selective absorber coatings is performed in a box-type SS-based vacuum chamber having 550 mm (width) × 450 mm (height) × 550 mm (length) dimensions. A heating arrangement is integrated with this chamber, which can be used up to 800°C at different heating rates. Digital Proportional Integral Derivative (PID) controller in conjunction with a K-type thermocouple is used to control/measure the heating system. This chamber has dynamic vacuum of pressure approximately 5×10^{-7} mbar, which can be monitored using a microprocessor-controlled Pirani gauge for pressure up to 10^{-3} mbar and using a Penning gauge from pressure 10^{-3} to 10^{-7} mbar.

3.3 SAMPLE CHARACTERISATION TECHNIQUES

After the design and development of spectrally selective coating structures, the physical, microstructural, optical, and solar thermal properties of these coatings need to be investigated. Numerous characterisation techniques are used to understand the physical properties such as structural, microstructural, elemental, optical, mechanical, thermal, corrosion, and thickness of spectrally selective absorber coatings. These techniques include X-ray diffraction (XRD), Raman vibrational spectroscopy, scanning electron microscopy (SEM), energy-dispersive X-ray (EDX) measurement, atomic force microscopy (AFM), UV-Vis and UV-Vis-NIR spectrophotometer, Fourier transform infrared (FTIR) spectrophotometer, nanoindentation through atomic force microscopy (AFM), cyclic voltammetry, and Tafel measurements. The basic principle of these techniques and description about the respective equipment, which is used to characterise these spectrally absorber/selective coatings, are discussed in the following section.

3.3.1 X-Ray Diffraction (XRD)

X-ray diffraction technique is an extremely powerful technique, used to identify the crystallinity and related crystal phases of the materials. Wilhelm Conrad Röntgen, a German physicist, invented X-rays in 1885, which are electromagnetic radiation within 0.01–100 A° wavelength ranges. Moreover, X-ray diffraction was first invented in 1912, and it was extensively used for structural characterisation, especially the crystallographic structure, crystallinity, and more specifically crystallite size. The X-rays, electromagnetic radiation of short wavelength, are produced during collision between the high-speed accelerated electrons and a metal target. The accelerated electrons decelerate in the metal target and, as a consequence, generate the continuous X-ray spectrum. In this process, some of the high-energy electrons knock out the innermost target electrons, generating characteristic X-rays, superimposed with continuous X-ray background.

The X-ray tube is used to generate X-rays and consists of a metal target and a thermionic electron source. The schematic of the X-ray tube is shown in Figure 3.10a. A high voltage is applied across the electrodes to accelerate the electrons towards the target (anode). An immensely cooling system is used to cool the X-ray tube. Two types of X-rays are generated, continuous and characteristic X-rays. X-ray diffraction techniques normally require monochromatic radiation, known as the characteristic

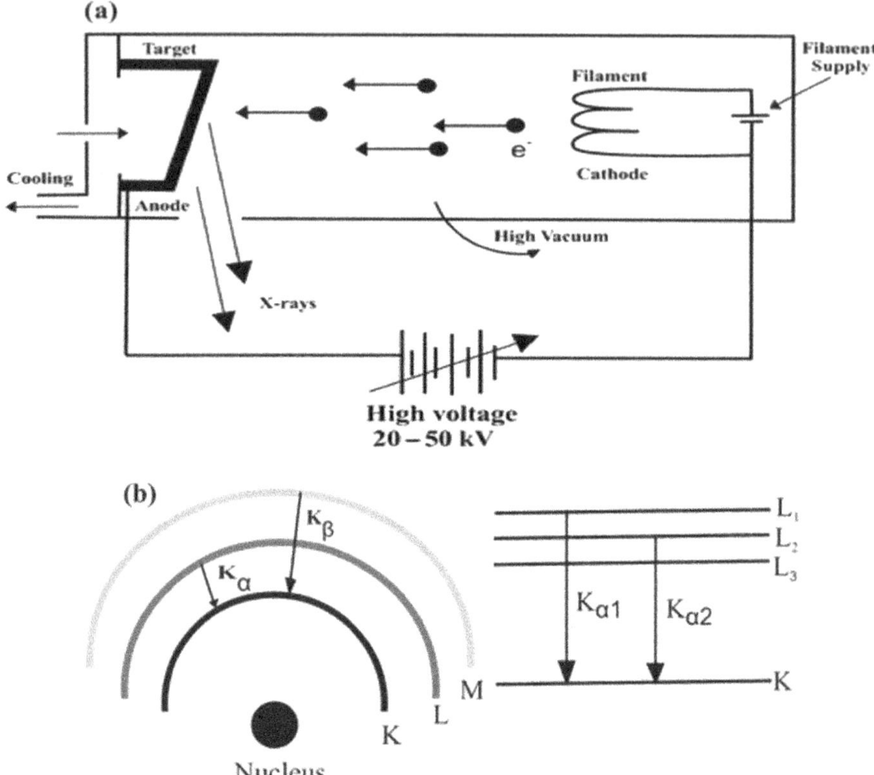

FIGURE 3.10 The schematic of (a) X-ray tube structure and (b) characteristic X-ray generation.

Source: Leng (2013)

X-rays. The schematic illustration for the characteristic X-ray generation is demonstrated in Figure 3.10(b), where innermost electrons of the target material are knocked out, and the created vacancy is filled with the higher orbital electron, releasing the characteristic photon of that energy difference.

X-ray diffraction techniques are based on wave interference. It can extract the information of inter-planar spacing using Bragg's law $\left(n\lambda = 2d\sin\theta\right)$. The organisation of the X-ray source, specimen, soller slit, monochromator, and detector is shown in Figure 3.11(a). The generated X-rays pass through the special soller slit, and, thus, a collimated X-ray beam is formed. This collimated X-ray beam through a divergence slit is allowed to fall on the specimen, and the diffracted beams are collected at the detector after passing through receiving slits, soller slit, and monochromator. The commonly used diffractometer is based on the Bragg–Brentano (B-B) geometry (Figure 3.11) and is suitable for thin films' analysis. An additional optical arrangement is used for generating a parallel incident beam in conjunction with B-B geometry to find out the phases in thin-film structures [Leng, 2013].

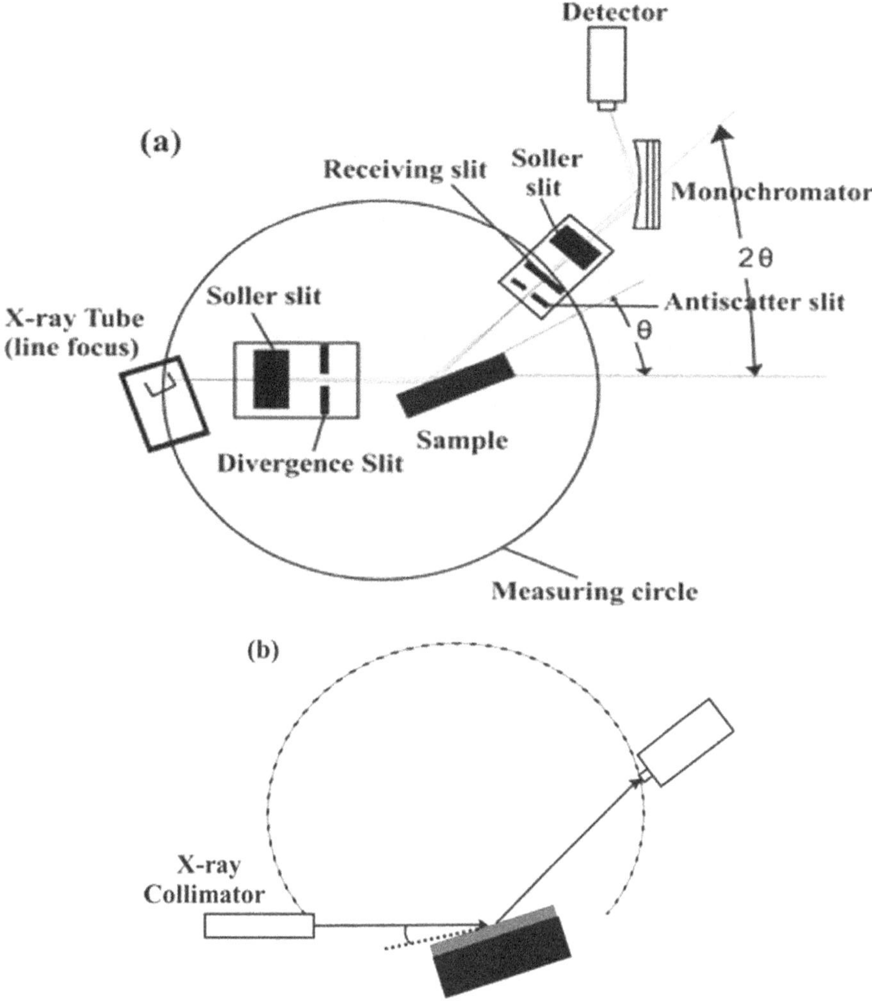

FIGURE 3.11 (a) Schematic of the geometrical arrangement of X-ray diffractometer and (b) schematic of the optical arrangement of thin-film diffractometer.

Source: Leng (2013)

In this work, we used D8 Advance Powder X-ray diffractometer in conjunction with parallel beam geometry, to carry out the structural analysis of the investigated spectrally selective absorber coatings. Figure 3.12 shows the D8 advance powder X-ray diffractometer system at IIT Jodhpur, used to characterise the structural properties in this study. The details of the parameters used at the time of the collecting X-ray diffraction data for the selective surface are explained in Results and Discussion sections in different chapters.

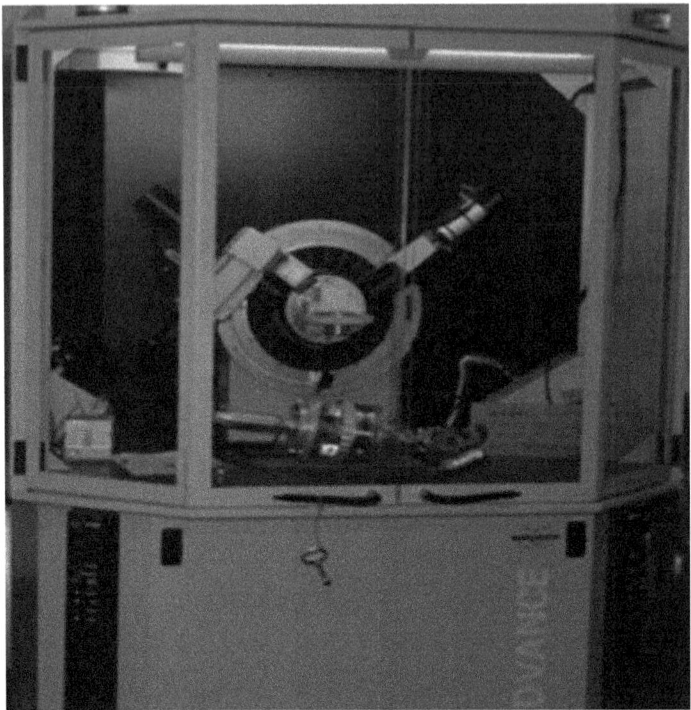

FIGURE 3.12 D8 advance powder X-ray diffractometer system in our lab used to characterise the structural properties in this study.

3.3.2 Raman Vibrational Spectroscopy (RVS)

RVS is used to examine the structure of molecules using inelastic interaction between incident monochromatic electromagnetic radiation and atomic vibrations in molecules and crystals. Vibrational Spectroscopy (VS) is usually operated in the range of infrared (~ 10^{-7} m), because of the probable energy range of the vibrational energy levels. VS identifies the molecular/crystal vibrational energy levels either by absorption and/or by inelastic scattering of the incident light through a molecule/crystal.

Raman spectroscopy (RS) is based on the inelastic Raman scattering phenomenon of incident monochromatic radiation by molecules/crystals. Monochromatic radiation will be scattered both elastically and inelastically after interacting with the molecule/crystal. Elastic scattering consists of a frequency identical to that of the incident one, whereas, in the case of inelastic scattering, the scattered lights have different frequencies from that of the incident one. Inelastic scattering, in specific cases, is called Raman Scattering. There are two types of inelastic scattering processes. In one case, the scattered light has more energy than that of incident light, and the process is called anti-Stokes process, whereas in other, the scattered light has less energy than that of incident light, and the process is called Stokes process.

The probability of anti-Stokes process is much lesser than Stokes process, and that is why Raman vibrational spectrums are recorded in Stokes mode usually. Moreover, the recorded intensity as a function of change in frequency or wave number is called Raman spectrum.

Molecules/crystals are made of atoms, which vibrate around their equilibrium positions. The vibrational frequencies depend on molecules/crystals' atomic bonding, surrounding environment, and crystallographic information. The vibrational modes are usually defined in terms of normal modes of the system, which depends on the total degrees of freedom available to the system. Among these vibrational modes, only modes, which exhibit a change in polarisability, are Raman active and are observed in Raman vibrational spectroscopic measurements.

Figure 3.13(a) illustrates the optical arrangement of a Raman vibrational spectrophotometer. A monochromatic light (mostly laser light) is focused on a sample surface using the optical microscope. The Raman scattered signals, produced due to the inelastic scattering, are too weak as compared with the Rayleigh scattered signals. Thus, a holographic filter has to be used to block the Rayleigh scattered signal and to allow only the Raman scattered signals. The wavelength dispersion of Raman scattering is selected by a diffraction grating system before being recorded at the detector.

Nomadic™ Raman Microscope (BaySpec, US make) has been used to analyse the vibrational properties of spectrally selective absorber coatings. Figure 3.13(b) shows the Nomadic Raman microscope at IIT Jodhpur, used to measure the vibrational properties of selective absorber materials in this study. Nomadic™ Raman microscope is equipped with multiple excitation sources (355 nm, 532 nm, and 785 nm), for Raman vibrational spectroscopy experiments. In the present study, a 532-nm green monochromatic incident light source has been used. The system also consists of a dedicated spectrograph and optimised detectors for individual wavelength source to ensure optical spectral coverage, resolution, and sensitivity. It is also featured with integrated laser control and motorised ND filters for attenuating laser power. A fully functional Olympus microscope is integrated with motorised stage, which provides a 3"× 2" platform for transmittance and reflectance mode operation. Moreover, a CCD camera is equipped on the top of the microscope for capturing bright field images. Details of the experiments are given in Results and Discussion section of Chapter 5.

3.3.3 Scanning Electron Microscope (SEM)

Scanning electron microscope (SEM) is used to probe the microscopic structure, more precisely the surface structure, by scanning across the surface. The essential components of SEM are an electron gun, series of electromagnetic lenses, and apertures. Figure 3.14(a) shows the schematic skeleton of an SEM. The electromagnetic lenses are used for collimated electron beam formation, and the objective lenses are used to focus the electron beam as a nanometre diameter probing electron beam. Probe scanning is carried out by using the beam-deflection system, integrated within the objective lens of the SEM. The deflection system of the

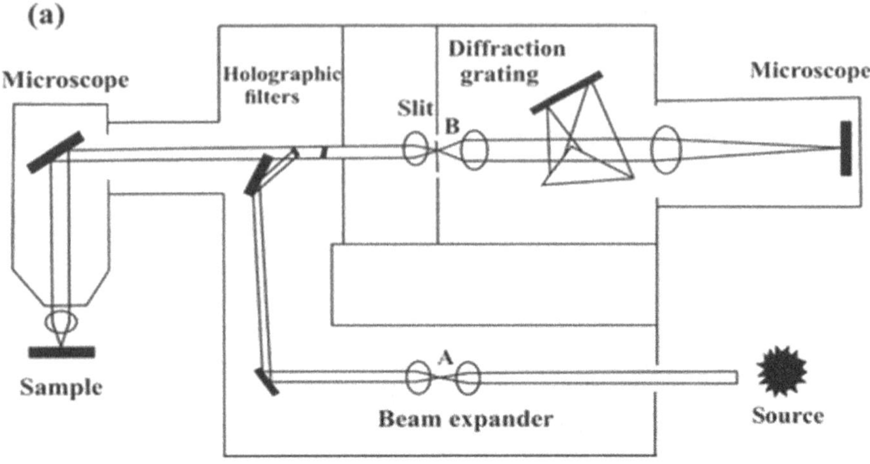

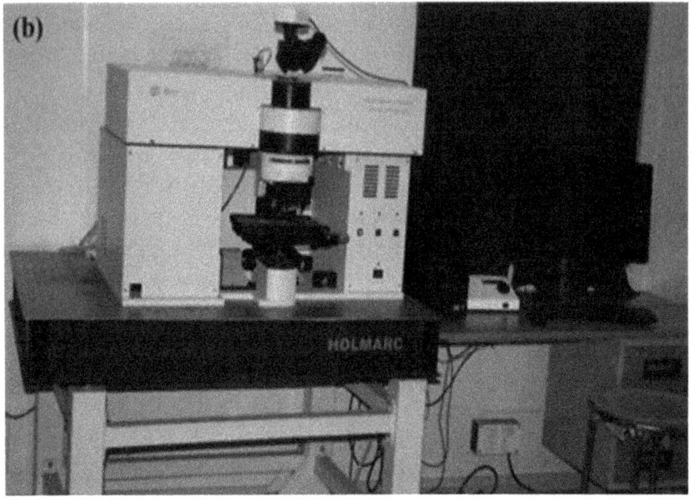

FIGURE 3.13 (a) Schematic of the optical diagram of Raman microscope (source: Leng, 2013). (b) Nomadic Raman microscope in our lab used to measure the vibrational properties of selective absorber materials in this study.

electron probe is controlled by pairs of electromagnetic coils (also called scanning coils). Apertures are mainly used for limiting the divergence of the electron beam in its optical path.

To obtain good SEM images, it is essential to control the operational parameters to get the desired depth of field (DoF) and resolution. The DoF is strongly affected by the working distance and aperture size. The resolution of the images is adjusted by the two operational variables: (i) the acceleration voltage of electron gun and (ii) probe current. Astigmatism is another operational variable and needs to be minimised for better microscopic information. These parameters become significant at

(a)

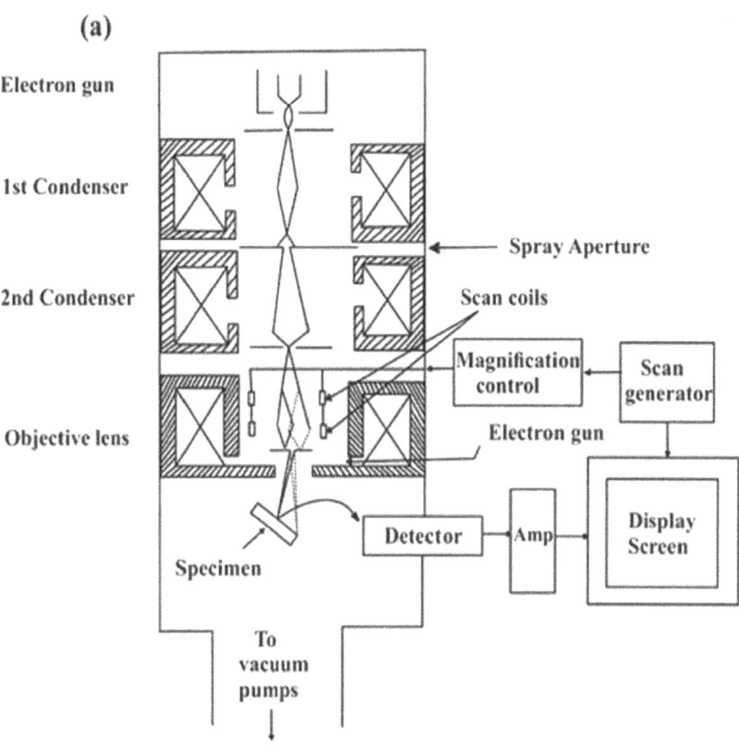

(b)

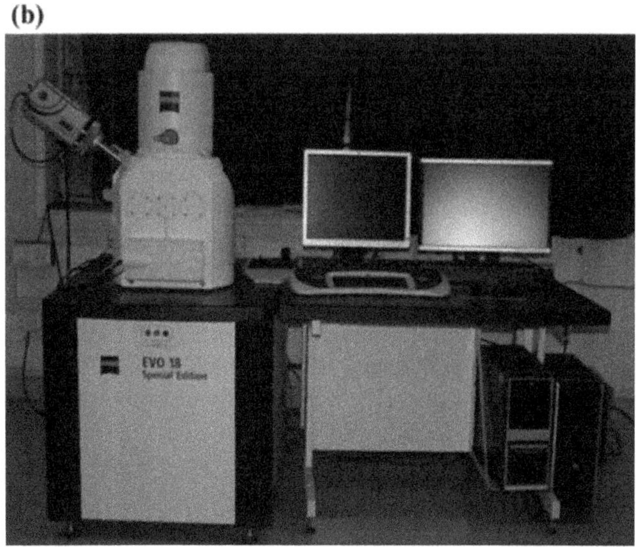

FIGURE 3.14 (a) Schematic of the structure of a scanning electron microscope (SEM) (source: Leng, 2013). (b) Carl Zeiss scanning electron microscope system in conjunction with energy-dispersive spectroscopy (EDS) in our lab used to characterise the surface properties and elemental analysis of selective absorber films.

higher resolution and need to be adjusted properly to get clear scanning electron microscopic images.

Carl Zeiss SEM EVO 18 special edition has been used to analyse the surface and cross-sectional morphologies and grain sizes of the spectrally selective absorber materials. Figure 3.14(b) shows a Carl Zeiss scanning electron microscope system in conjunction with an energy-dispersive spectroscopy (EDS) accessory at IIT Jodhpur, used to characterise the surface properties and elemental analysis of selective absorber films.

3.3.4 X-RAY ENERGY-DISPERSIVE SPECTROSCOPY (EDS)

EDS is an X-ray spectroscopy technique. It uses characteristic X-rays to identify chemical elements in a material. EDS is generally an add-on component with electron microscopes such as SEMs and transmission electron microscope (TEMs). The EDS instrument [Oxford make (X-act)] is attached with Carl Zeiss SEM EVO 18 special edition, as shown in Figure 3.14(b). This is used to evaluate the elemental compositions in spectrally selective absorber material coatings used in this thesis work.

3.3.5 ATOMIC FORCE MICROSCOPE (AFM)

An AFM is a popular type of scanning probe microscopy (SPM), which was invented in late 1980s. An AFM uses near-field forces between atoms of the probe tip apex and surface to generate signals for surface topography, and Figure 3.15(a) shows the schematics of AFM. A probe tip, which is mounted on a cantilever spring, is used to scan the surface. Near-field forces between the tip and the sample surface are detected using a beam-deflection system, which senses the respective spacing. The main part of AFM, which differentiates it from other SPM members, is its microscopic force sensor. Normally, AFM is operated in two modes: (i) static and (ii) dynamic. Static mode must be operated in a contact type, whereas dynamic modes may be operated in both contact and non-contact types [Leng, 2013; Wiesendanger, 1994; Zhang *et al.*, 2009; Butt *et al.*, *2005*].

A Park System, scanning probe microscopy (SPM), XE-70 has been used to analyse the surface morphologies, surface roughness, grain size, and in some cases the thickness of spectrally selective absorber coatings. Figure 3.15(b) shows scanning probe microscopy (SPM) instruments at IIT Jodhpur, used to analyse surface parameters in this study. Details of the AFM experiment will be presented in Results and Discussion sections wherever AFM results are presented.

3.3.6 PROFILOMETER (THICKNESS MEASUREMENTS)

A profilometer is a device used to measure the roughness of the surfaces. It provides the relative difference between a high and low point on a surface in the range of nanometres to micrometres. A stylus (sharp tip)-based profilometer traces the surface and records the tip position using optical or electrochemical methods [Conroy and

(b)

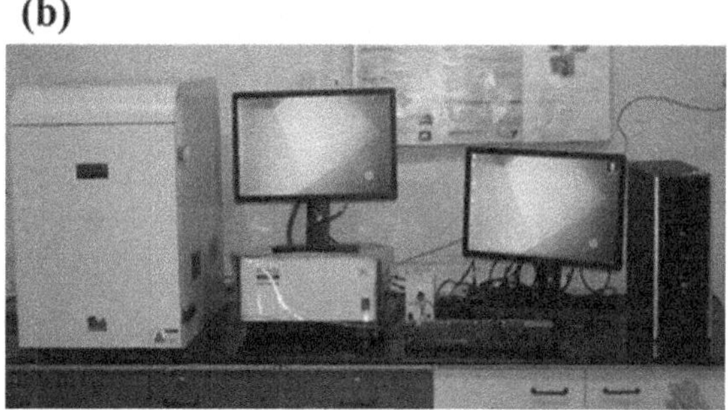

FIGURE 3.15 (a) Schematic of an atomic force microscopy (AFM) (source: Butt *et al.*, 2005). (b) Scanning probe microscopy (SPM) instruments in our lab used to analyse surface parameters in this study.

Amstrong, 2006]. A profilometer contains a stylus, which is in contact with the surface of samples and an electrochemical transducer that converts its Z coordinate into voltage. An amplifier and analogue-to-digital converter are connected to a computer to record measurements [Leach, 2011]. DektakXT stylus profiler has been used to measure the thickness of the spectrally selective films, coated on different substrates (Figure 3.16).

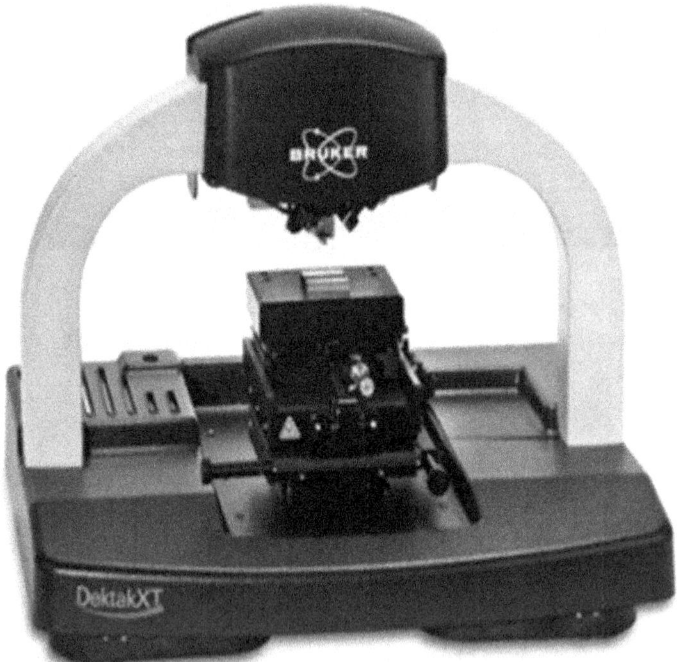

FIGURE 3.16 DektakXT stylus profiler at IIT Jodhpur, used to measure the film thickness of spectrally selective absorber coatings in the present study.

3.3.7 Optical Measurements

The optical properties of spectrally selective absorbers are commonly examined using spectrophotometers in the desired wavelength ranges. UV-Vis-NIR spectrophotometers are of special interest, which can cover the entire solar spectrum $(0.2–2.5\,\mu m)$. In contrast, UV-Vis spectrophotometers cover around 50% of the wavelength range of the solar spectrum (0.2–0.9 μm) only and completely exclude the infrared part of the solar spectrum. Thus, it is important to use UV-Vis-NIR spectrophotometer to understand the complete spectral response of spectrally selective coatings.

Fouriertransforminfrared(FTIR)spectrophotometerscovertheinfrared(2.5–25μm) region. This range is important to understand the materials' optical/thermal properties, especially emissivity, and may be considered feedback for the development of desired physical properties. In this section, the basic principle of spectrophotometer techniques, used to measure the optical properties (absorptance and emittance) of selective absorber materials, will be discussed.

3.3.7.1 UV-Vis-NIR/UV-Vis Spectrophotometer

Agilent Technology Cary 5000 UV-Vis-NIR spectrophotometer is used to collect the total and diffuse reflectance, which covers 0.175 – 3.3 μm of the wavelength range, thus the entire solar spectral range. This UV-Vis-NIR spectrophotometer is a double-beam system, equipped with a photomultiplier detector for the UV-Vis range and PbS detector for NIR range. A 150-mm diameter integrating sphere accessory, an external DRA, has been used in normal scan mode to get the reflectance in the desired wavelength range. The external DRA system can be combined with the main instrument. The integrating sphere is an optical device used to collect and measure electromagnetic radiation from an object. The integrating sphere is coated with a 4-mm thick polytetrafluoroethylene (PTFE), a white diffusive material and fitted in the same detector configuration as the main spectrophotometer during diffuse reflectance experiments. The reflectivity of the PTFE is above 96% between 0.2 and 2.5 μm, and greater than 99% between 0.3 and 1.8 μm. Figure 3.17(a) shows the schematic view of external DRA and optical design view of the external DRA (b), used to measure the total reflectance of spectrally selective absorbers for the present work.

In DRA measurements, a baseline is collected first using the PTFE reference disk, and then the sample is mounted over the sample reflectance port and the diffused reflection of the sample is collected by the integrating sphere. The solar absorptance, α, of the selective absorbers, is calculated by measuring the total reflectance in the solar wavelength range and weighing by 1.5 AM solar radiation using Equation (2.5).

In this study, Cary 4000 UV-Vis spectrophotometer has also been used to collect the total reflectance of selective absorbers in the range of 0.2–0.9 μm. This equipment has only the photomultiplier (PMT) detector and in NIR region it is not able to detect. This spectrophotometer is also equipped with 110-mm-diameter integrating sphere for DRA measurements. The integrating sphere can be easily installed using the lock down mechanism in the sample compartment of the instrument. Figure 3.18 represents the schematic view of the optical design of the internal DRA (a) and beam diagram when the sphere is kept in the D (b) and S (c) positions of the accessory. If accessory is kept in the D position, the detector only detects the diffuse reflectance. However, in S position, the detector can collect the total reflectance, including the specular one as well.

3.3.7.2 Fourier Transform Infrared Spectrophotometer (FTIR)

FTIR has extensively been used for vibrational spectroscopic measurements. It depends on the interaction of infrared radiation with the vibrating dipole moments

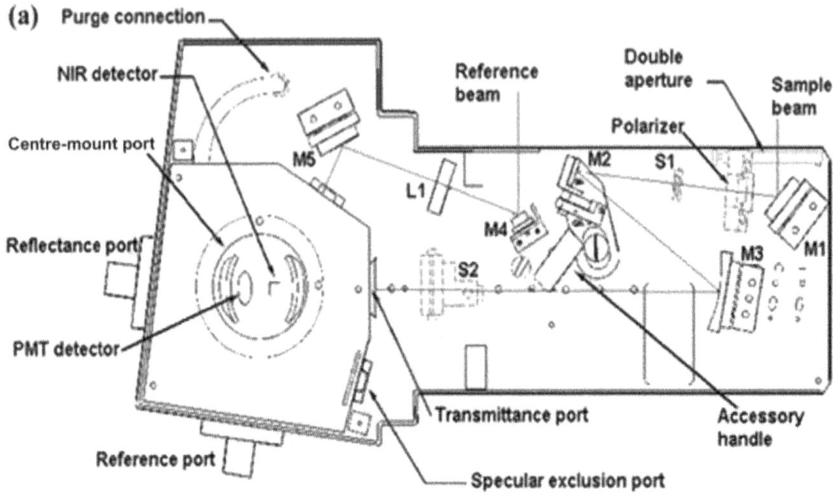

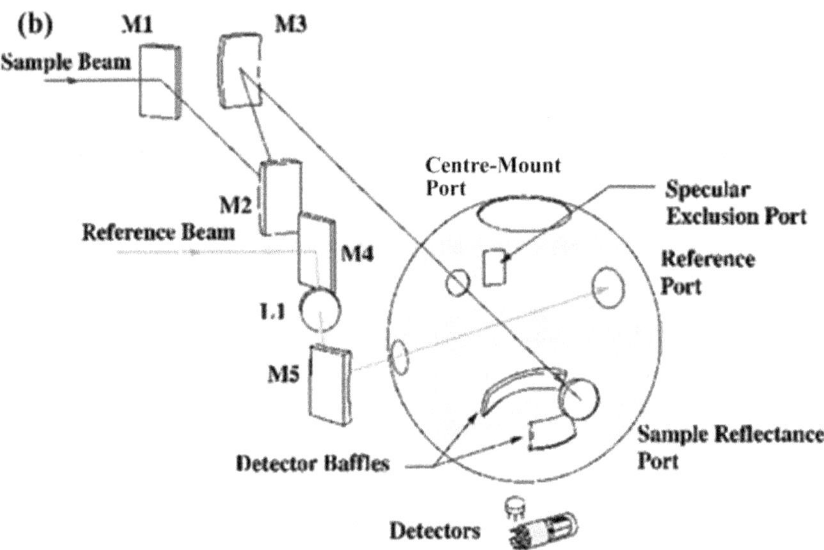

FIGURE 3.17 (a) A schematic view of the external DRA and (b) the optical design view of the external DRA used to measure the total reflectance of spectrally selective absorbers in this study.

Source: Manual of Cary 5000 VU-Vis-NIR, Agilent Technology

of molecules. Reflectance spectra of the spectrally selective absorbers are recorded in 2.5–25 μm wavelength range to understand the infrared/thermal response of these coatings. We used a Bruker FTIR spectrometer vertex 70v, and snap of the same is shown Figure 3.19. This spectrometer consists of a room-temperature Deuterated and L-alanine Doped Triglycine Sulphate (DLaTGS) detector. The resolution of this FTIR spectrophotometer is approximately 0.4 cm^{-1}.

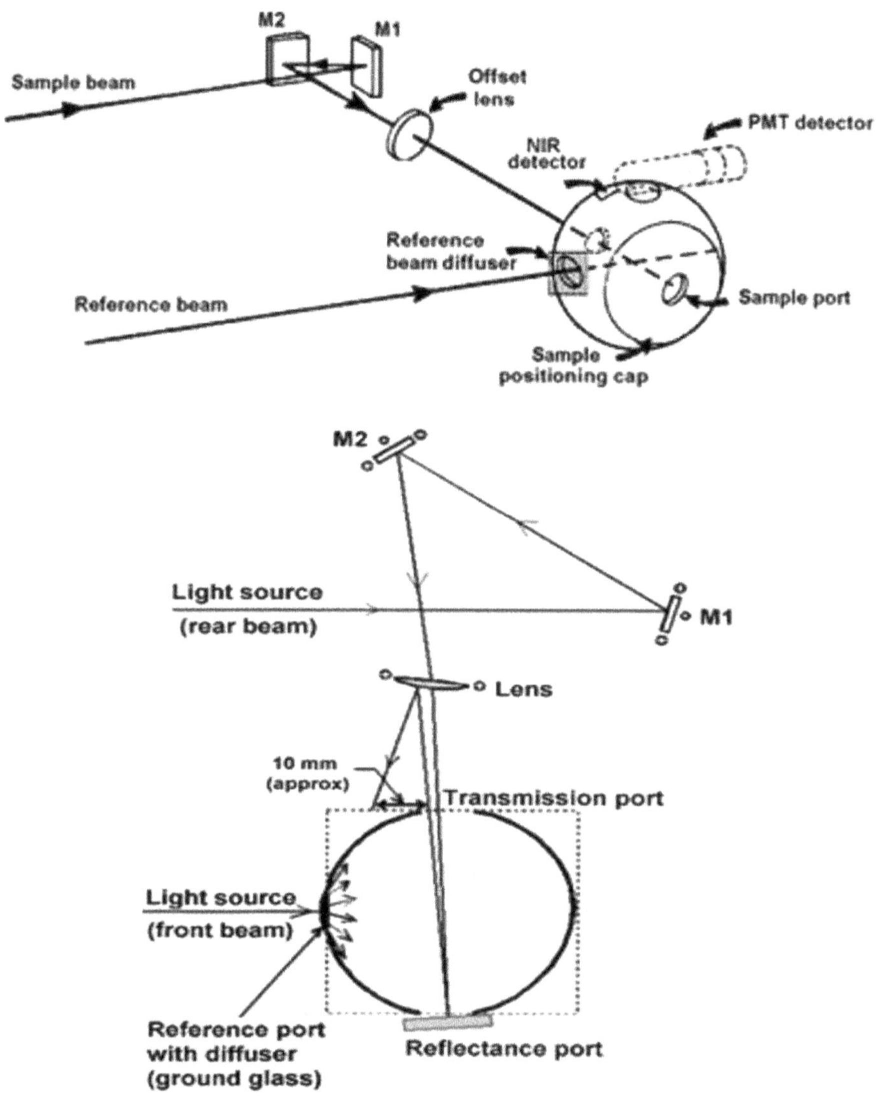

FIGURE 3.18 A schematic view of the optical design of internal DRA (a), beam diagram of the accessory with the sphere cap in the D position (b), and beam diagram of the accessory with the sphere cap in the S position (c).

Source: Manual of Cary 4000 VU-Vis spectrophotometer, Agilent Technology

FTIR 70v spectrometer has a fully evacuated optical bench, which is used to remove the residual environmental background such as gases and moistures during the experiments. A 30-degree fixed-angle specular reflectance accessory, as shown in Figure 3.20, has been used to collect the reflectance spectra of the spectrally selective absorbers. The sample is placed on the hole in the attached platform, as explained in Figure 3.20. The background spectrum is collected by using a gold-coated reference

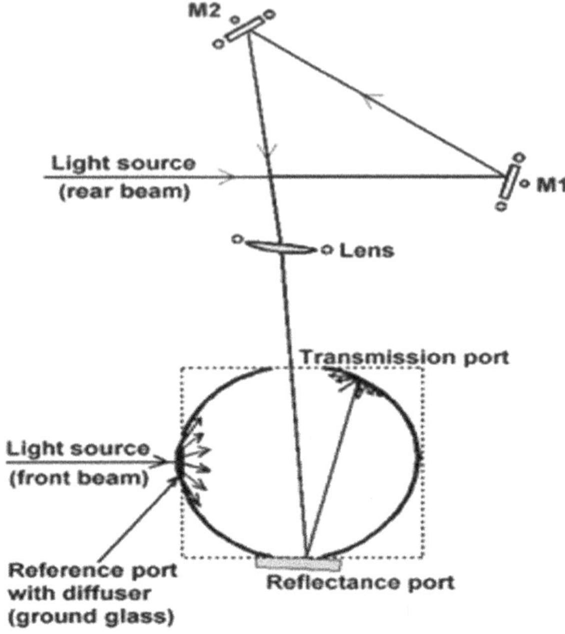

FIGURE 3.18 (Continued)

FIGURE 3.19 Bruker FTIR spectrometer vertex 70v in our lab used to collect the reflectance spectra of spectrally selective absorbers.

sample, as an ideal reflector, for the infrared region. The FTIR instrument is used to collect the total specular reflectance in 2.5–25 μm wavelength range, and the thermal emittance has been calculated from these measurements using Equation (2.7).

3.3.8 Thermal Analysis

Thermal analysis is an analytical technique that measures the properties or changes in properties of materials as a function of temperature. There are several common

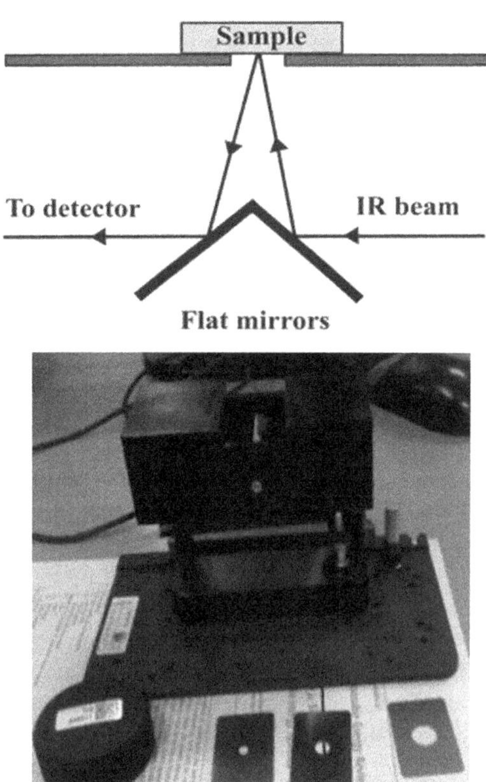

FIGURE 3.20 (a) Optical diagram of fixed-angle (30 degrees) specular reflectance accessory (source: Leng, 2013). (b) Thirty-degree horizontal fixed-angle specular reflectance accessory.

thermal analysis techniques such as thermogravimetry (TG), differential thermal analysis (DTA), and differential scanning calorimetry (DSC). TG is particularly employed to check the decomposition of materials by invigilating the mass change as a function of temperature. DTA and DSC are widely used to analyse the phase change as a function of temperature for different materials [Leng, 2013]. Figure 3.21 shows the schematic diagram of a thermal analysis instrumentation, which consists of the sample with control heating option in conjunction with a high-precision weighing balance. The systems are usually automated/interfaced with a computer for data logging and analysis.

The key component of the TG instrument is the microbalance, which measures the change in mass of the sample. Figure 3.22(a) shows the schematic of a null-type microbalance (commonly used), also called the Cahn microbalance. TG analysis mostly relies on the mass of the sample. Normally, a small amount of the sample is better with respect to the large amount to minimise the temperature inhomogeneities across the sample during the measurements. Moreover, the mass of the sample is usually about several milligrams. TG measurements can be done in both reactive and non-reactive atmospheres. The non-reactive atmosphere should be created using

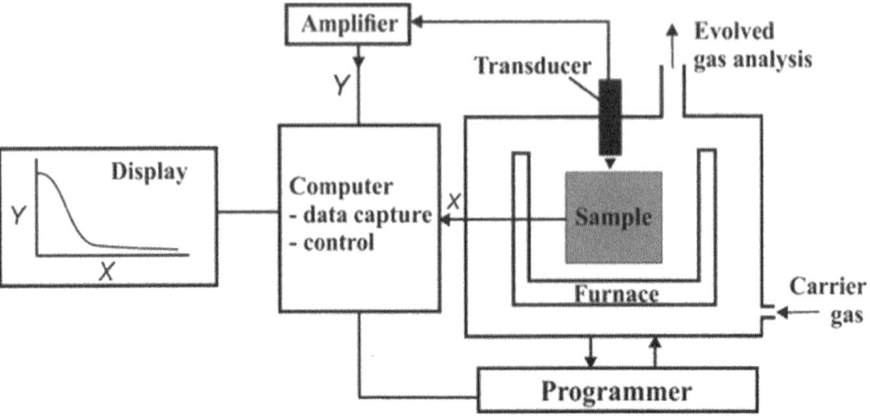

FIGURE 3.21 Schematic of the general instrumentation of thermal analysis.

Source: Leng (2013)

an inert gas. Normally, argon (Ar) and nitrogen (N_2) gases are used for creating non-reactive atmospheric conditions. TGA is a simple and effective technique to estimate the thermal stability and chemical reactions by monitoring mass change for materials of interest [Leng, 2013]. In order to measure the thermal stability of spectrally selective absorber materials, the simultaneous thermal analyser (STA) is used. STA 6000 is used to measure TG responses for different coating structures. Figure 3.22(b) shows a Perkin Elmer simultaneous thermal analyser 6000 (STA 6000) at IIT Jodhpur, used for thermal gravimetric measurements of spectrally selective absorber films. The instrument is equipped with a top-loaded furnace, where temperature can be varied from 15°C to 1,000°C at different heating scan rates from $0.1\ \text{to}\ 100°\text{C}/\text{min}$ with sample capacity of approximately 1,500 mg. Balance sensitivity of 6000 STA is of the order of 0.1 μg.

3.3.9 Mechanical Characterisation

Mechanical properties, such as Young's modulus (YM) and hardness, of spectrally selective absorbers are measured using nanoindentation method. The nanoindenter, which is used in this study, is an integrated accessory with atomic force microscopy (scanning probe microscopy (SPM) XE-70, Park) system. In this study, the indenter, used for measurement, is a Berkovich indenter and made of a sapphire cantilever attached with a diamond tip. Tip stiffness, radius, height, thickness, and length are $140\,\text{N/m}$, $< 25\,\text{nm}$, $96\,\mu\text{m}, 26\,\mu\text{m}$, and $717\,\mu\text{m}$, respectively. The respective frequency, half angle, front angle, side angle, inclination values for the tip are 52 kHz, 30°, 90°, 79°, $12°$, respectively. Oliver and Pharr analysis has been used to calculate the hardness and YM of spectrally selective absorbers [Oliver and Pharr, 1992, 2004]. The key parameters, used to analyse the hardness and Young's modulus, are the maximum load, *Pmax*, the maximum displacement, *hmax*, and the elastic unloading stiffness, S = dP/dh, defined as the slope of the upper portion of

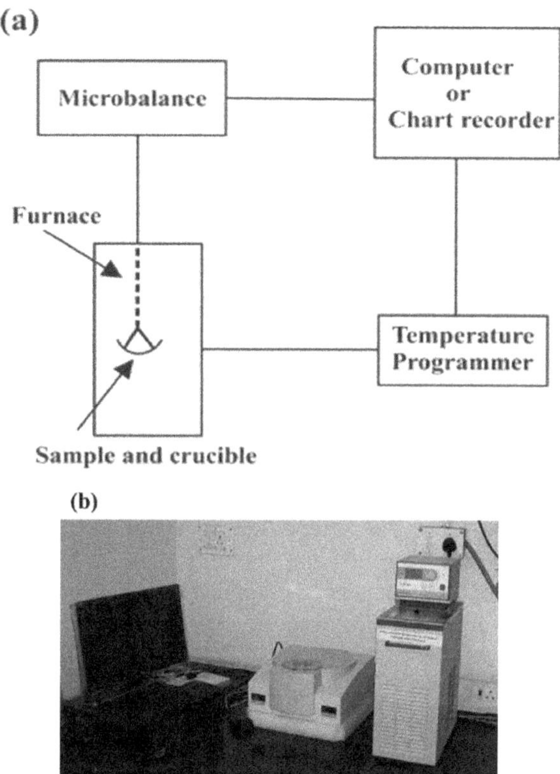

FIGURE 3.22 (a) Schematic of commonly used null-point type microbalance (source: Leng, 2013). (b) Perkin Elmer simultaneous thermal analyser 6000 (STA 6000) in our lab used for thermal annealing of spectrally selective absorber films.

the unloading curve during the initial stages of unloading. From the analysed load–displacement curves, Young's modulus can be calculated using Equation (3.4).

$$\frac{1}{E_{eff}} = \frac{2\beta}{S}\sqrt{\frac{A}{\pi}} = \frac{1-v_s^2}{E_s} + \frac{1-v_i^2}{E_i}, \quad \text{Eq. (3.4)}$$

where A, E_{eff}, S, and β are the actual contact area, effective (or reduce) modulus for each indenter/specimen combination, measured stiffness, and a shape constant (β) which is 1.034 for Berkovich tip [Fang *et al.*, 2004]. E and v represent Young's modulus and Poisson's ratio of the indenter (i) and the sample (s), respectively. The values used for Poisson's ratio v and Young's modulus Ei for diamond indenter are 0.07 and 1141 GPa [Oliver and Pharr, 1992]. Hardness values can be calculated using Equation (3.5) [Oliver and Pharr, 1992, 2004], where A is the effective area of contact.

$$H = \frac{P_{max}}{A} \quad \text{Eq. (3.5)}$$

The details of the nanoindentation measurement parameters in this study are discussed in respective chapters.

3.3.10 Electrochemical Measurements

Electrochemical measurements, such as cyclic voltammetry (CV) and linear sweep voltammetry (LSV), are used to understand the chemical response of a system under an electrical excitation (stimulation). This section covers CV and LSV electrochemical techniques, which are used to understand the oxidation, reduction and corrosion response of spectrally selective absorbers in this study.

3.3.10.1 Cyclic Voltammetry (CV)

CV is an extensively used electroanalytical technique for electrochemical measurements. In this technique, the potential difference between working and the reference electrode is swept linearly in time from a start potential to a final potential and reversed to complete a cycle. The current response of a small stationary electrode in an unstirred solution is excited by a triangular potential waveform (Figure 3.23(a)). The waveform produces the forward and reverse scans. The resulting current at the working electrode is collected against the applied electrode potential in voltammogram. Figure 3.23(b) illustrates the respective response of a reversible redox couple during a single potential cycle [Wang, 2006; Skoog *et al.*, 2004; Kissinger and Heineman, 1996a].

The mostly used experimental configuration for collecting CV consists of an electrochemical cell system (Figure 3.24(a)), with three electrodes: (i) counter or auxiliary, (ii) reference electrode (RE), and (iii) working electrode (WE). These electrodes are dipped in a liquid electrolyte and connected to a potentiostat. A potentiostat is used to measure the potential difference between the reference and working electrodes. A counter electrode with this configuration is used for making accurate current measurements between the WE and RE. A counter electrode plays an important role to ensure that no current is passing through the reference electrode, and the potential across the reference electrode remains constant during the measurement.

In this study, Iviumstat electrochemical workstation (Figure 3.24(b)) has been used to measure CV response for spectrally selective absorber films coated on several substrates such as copper, stainless steel, and aluminium. Ag/AgCl electrode and platinum wire electrode are used as reference and counter electrodes. The selective absorber film-coated substrates (Cu, SS, Al) are used as working electrode. NaCl saline solutions (3.5 wt.%) are used as the electrolyte for evaluation of corrosion properties. To understand the CV response of interested structures, the experiments are carried out at several scan rates and several cycles, and the details of such experiments and related results are discussed in respective sections.

3.3.10.2 Corrosion Measurements

Corrosion is the degradation of materials by means of chemical reactions under environmental conditions, in which the material is used for possible applications. Linear sweep voltammetry is used for corrosion studies, in which potentiodynamic polarisation measurements are carried out to measure the corrosion of the spectrally selective

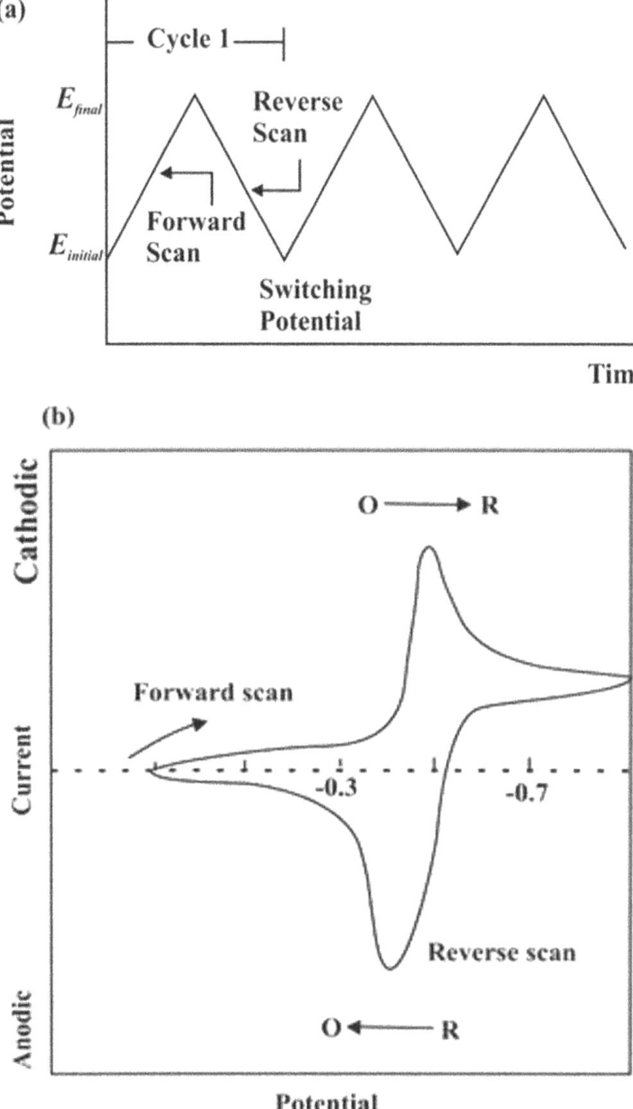

FIGURE 3.23 (a) Potential–time excitation signal in a cyclic voltammetric experiment. (b) A typical cyclic voltammogram for a reversible redox process.

Source: Wang (2006)

absorber films coated on the substrate. In a linear sweep measurement, potential across the working electrode is varied at a constant rate throughout the scan, and the resulting current is measured.

To perform the polarisation measurements, an electrochemical three-electrode (working, counter, and reference) test setup, integrated with a potentiostat, has been

(a)

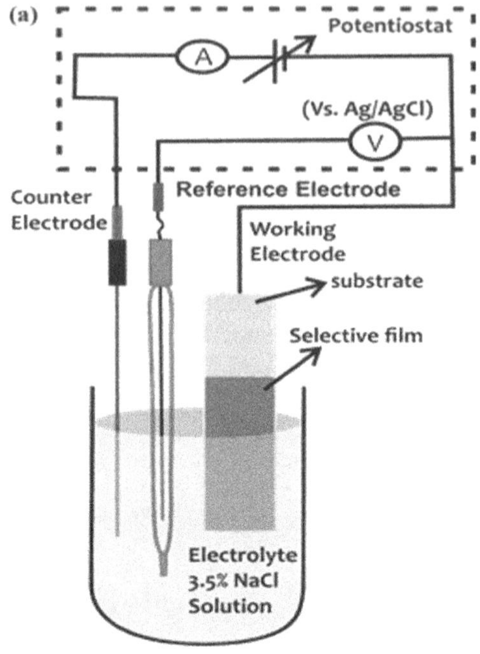

(b)

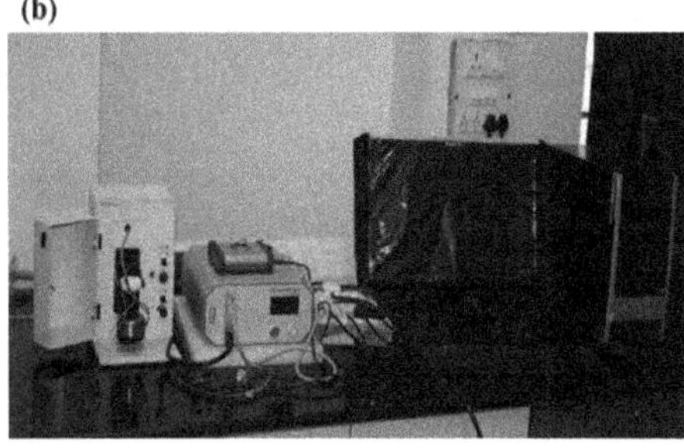

FIGURE 3.24 (a) Schematic of the experimental setup for cyclic voltammetry experiments with a three-electrode configuration. (b) Iviumstat electrochemical workstation in our lab, used for the electrochemical measurements of spectrally selective absorber coatings.

designed and used for corrosion experiments (Figure 3.24(b)). Working electrodes, whose potential is varied linearly with time, versus a reference electrode, are spectrally selective structures in the present studies. The potential at the reference electrode is kept constant throughout the experiments. The current is measured between the WE and CE. These measurements are used for Tafel analysis to estimate the corrosion resistance $\left(R_P\right)$, corrosion current density $\left(i_{corr}\right)$, and corrosion potential $\left(E_{corr}\right)$ relative to reference electrode by extrapolating the straight-line portions of the anodic and cathodic lines.

The corrosion currents are determined using Tafel plot analysis to understand the corrosion rates of these spectrally selective coatings. The Butler–Volmer (B-V) equation (Equation 3.6) gives the relation between the current density i and the charge transfer overpotential η, in terms of exchange current density i_0 and transfer coefficient α [Paunovic and Schlesinger, 2006].

$$i = i_0\left[\exp\left(\frac{(1-\alpha)zF\eta}{RT}\right) - \exp\left(-\frac{\alpha zF\eta}{RT}\right)\right], \qquad \text{Eq. (3.6)}$$

where $F = eN$ *is* Faraday constant, z is the number of electrons, R is the gas constant, and T is the absolute temperature. When an electrode is a part of an electrochemical cell through which current is flowing, the overpotential η is defined as the difference between applied potential and corrosion potential *Ecorr*.

$$\eta = E - E_{corr} \qquad \text{Eq. (3.7)}$$

Ecorr is open circuit potential of the corroding materials and also termed as corrosion potential.

For a large anodic current (η has large values), the B-V equation simplifies to the Tafel equation for the anodic reaction (Equation 3.8):

$$\eta = \log i_{corr} + b_a \log i \qquad \text{Eq. (3.8)}$$

Where b_a is the anodic Tafel slope.

$$\log i_{corr} = -\frac{(a(1-\alpha)zF)}{2.303RT}; b_a = \frac{2.303RT}{(1-\alpha)zF} \qquad \text{Eq. (3.9)}$$

For a large cathodic current (η has large values), the B-V equation simplifies to the Tafel equation for the cathodic reaction (Equation 3.10):

$$\eta = \log i_{corr} - b_c \log|i| \qquad \text{Eq. (3.10)}$$

Where *bc* is the cathodic Tafel slope.

$$\log i_{corr} = -\frac{a\alpha zF}{2.303RT}; b_c = \frac{2.303RT}{\alpha zF} \qquad \text{Eq. (3.11)}$$

The corrosion current i_{corr}, anodic Tafel slope b_a, and cathodic Tafel slope b_c can be measured from the experimental data. Stern–Geary equation [Stern and Geary, 1957] can be used to calculate corrosion current density, i_{corr}, which is as follows:

$$i_{corr} = \frac{b_a b_c}{2.3R_p(b_a + b_c)} \qquad \text{Eq. (3.12)}$$

Where *ba* and *bc* are Tafel slopes, and *Rp* is the polarisation resistance. Finally, corrosion rate can be calculated using the following equation:

$$\text{Corrosion Rate (mpy)} = \frac{0.13\ \mathrm{i_{corr}}\ (\mathrm{E.W.})}{d} \qquad \text{Eq. (3.13)}$$

Where E.W. is equivalent weight of the corroding material in gram, *d* is the density of the corroding material in g / cm^3, *icorr* is the corrosion density in $\mu A/cm^2$, and mpy is mils per year or mils penetration per year [Fontana, 2009].

In this study, Tafel analysis is used to measure the corrosion response of spectrally selective absorber films coated on several substrates. Schematic setup of corrosion measurements is illustrated in Figure 3.24(a) with Figure 3.24(b) showing the electrochemical experimental setup used to perform the corrosion measurements in this study. The details about corrosion behaviour of different spectrally selective structures are discussed in respective sections.

4 Low-Temperature Solar Selective Absorber Surface

Black Chrome-Graphite Encapsulated FeCo NPs Composite Structure

4.1 INTRODUCTION

Black chrome ($Cr-Cr_2O_3$) is one of the extensively studied and a commercialised spectrally selective absorber coating for photothermal applications. The historical development of black chrome (BC) spectrally selective absorbers has already been reviewed in Section 2.4.2. Several deposition techniques are used to deposit this spectrally selective absorber coating, explained briefly in Section 2.3. Among them, electrodeposition is one of the widely used deposition techniques due to its simplicity, low cost, and scalability towards large and curved surfaces [Quintana and Sebastian, 1994]. This deposition process is also suitable for the development of coatings with high solar absorptance, good stability in a wide range of oxidising/reducing environments, and high thermal resistance [Hamid, 2009]. Several studies are reported on the development of the electrodeposited black chrome coatings on numerous substrates like stainless steel (SS), copper (Cu), and bright nickel-coated Cu. Some of such black chrome spectrally selective coatings have been commercialised for solar thermal applications [Eugénio *et al.*, 2011; Duffie and Beckman, 1991; Schlesinger and Paunovic, 2000; Endres *et al.*, 2003, 2008; Endres, 2002; Zein El Abedin *et al.*, 2004; Mukhopadhyay *et al.*, 2005; Ali *et al.*, 1997].

The thickness of these coatings is very important and has an impact on solar thermal properties, especially the environmental impacts such as corrosion and related degradation. For example, the low thickness of such coatings cannot protect the substrate against atmospheric corrosion and thermal oxidation. A bright nickel coating, prior to the black chrome deposition, is recommended to solve the substrate corrosion problem to a small extent [Daryabegy and Mahmoodpoor, 2006]. The electrochemical characterisation of black chrome coatings, corrosion analysis of black chrome coatings, thermal behaviour, and the associated decomposition mechanisms and pathways have been studied in detail by several research groups across the globe [Abbott *et al.*, 2004; Ithurbide *et al.*, 2007; Al-Kuhaili and Durrani,

DOI: 10.1201/9781003563990-4

2007; Perissi *et al.*, 2006; Danilov *et al.*, 2001; Song and Chin, 2002; Lampert and Washburn, 1979; Sweet *et al.*, 1984; Inal *et al.*, 1981; Mabon *et al.*, 1982; Driver and McCormick, 1982; Surviliene *et al.*, 1999; Zajac *et al.*, 1980]. In order to understand the effect of corrosion on the optical properties of black chrome coatings, especially absorptivity (α) and emissivity (ε) were measured before and after corrosion tests in marine environment [Dibari and Turillon, 1979]. The thermal degradation of black chrome and the correlation of oxide content in black chrome coatings and their optical properties have been intensively studied and reported extensively in the literature [Grimmer and Collier, 1981; Hogg and Smith, 1977; Holloway *et al.*, 1980; Ignatiev *et al.*, 1979a, 1979b]. In conjunction with optical properties, other physical properties, such as mechanical properties, thermal stability (in air and vacuum), corrosion analysis, and its effects on solar thermal properties of electrodeposited black chrome coatings are studied and reported in the literature [Murr *et al.*, 1980; Prasad *et al.*, 1980; Rajagopalan *et al.*,1978; Reiss, 1981; Ritchie *et al.*, 1979; Spitz *et al.*, 1979; Valayapetre *et al.*, 1979; Window *et al.*, 1979; Zajac and Ignatiev, 1979; McDonald, 1975; Jafari and Rozati, 2011]. Nevertheless, studies on improving the corrosion and thermal stability of black chrome coatings are always attracting attention for enhanced solar thermal performance in ambient conditions. We have attempted the same, introducing nanoparticles in the black chrome selective coatings, which has not been explored much. Therefore, an attempt has been made to enhance the thermal stability using high temperature and environmentally stable graphite encapsulated FeCo nanoparticles (FeCo (C) NPs) into black chrome coatings for their application at or above 250°C temperature.

4.2 BLACK CHROME-GRAPHITE ENCAPSULATED FeCo NPs COMPOSITE SPECTRALLY SELECTIVE COATINGS

The thermal stability of black chrome spectrally selective coatings is still a challenge for their application at or above 250°C. This is mainly due to the oxidation of metallic chromium in black chrome coatings at an elevated temperature. The diffusion of environmental oxygen is responsible for such degradation, where oxygen reacts chemically with metallic chromium, converting it into chromium oxide. Thus, it is important to protect the oxidation of metallic content for its application at higher temperatures. Graphite encapsulated FeCo nanoparticles (FeCo (C) NPs) are of particular interest because of their high temperature and chemical/environmental stability, and that is why they are selected for the present study [Seo *et al.*, 2006; Turgut *et al.*, 1998; Desvaux *et al.*, 2005; Lee *et al.*, 2011; Poddar *et al.*, 2004; Chu *et al.*, 1999; Xu *et al.*, 2013]. The average particle size of FeCo (C) NPs, used in this work, is in the range of approximately 30 to 35 nm, with approximately 25 nm diameter of FeCo alloy as core and approximately 10-nm-thick onion-type lattice fringes of graphite carbon as shell [Gupta *et al.*, 2014]. These nanoparticles can be used to protect the metallic chromium nanostructures and thus may avoid the thermal oxidation of metallic chromium under ambient and elevated temperature conditions. A schematic representation of FeCo (C) NPs encapsulated metallic chromium content is shown in Figure 4.1. The upper part of the schematic, Figure 4.1, represents the metal chromium particles inside the chromium oxide (Cr_2O_3) matrix, forming a black chrome solar selective film on a copper substrate. The lower part of the schematic, Figure 4.1, explains the formation of FeCo(C) nanoparticle-protecting layer around the metallic chromium, inside the chromium oxide matrix, in nanoparticles modified black chrome spectrally selective coatings on a copper substrate. The middle portion of Figure 4.1 represents the zoomed region of FeCo (C) NP-covered Cr nanostructured metallic components. Thus, the high temperature and environment/chemical stability of FeCo (C) NPs may protect the oxidation of chromium metallic content. This may be the reason of observed higher thermal and corrosion stabilities of such FeCo (C) NPs modified black chrome spectrally selective coatings.

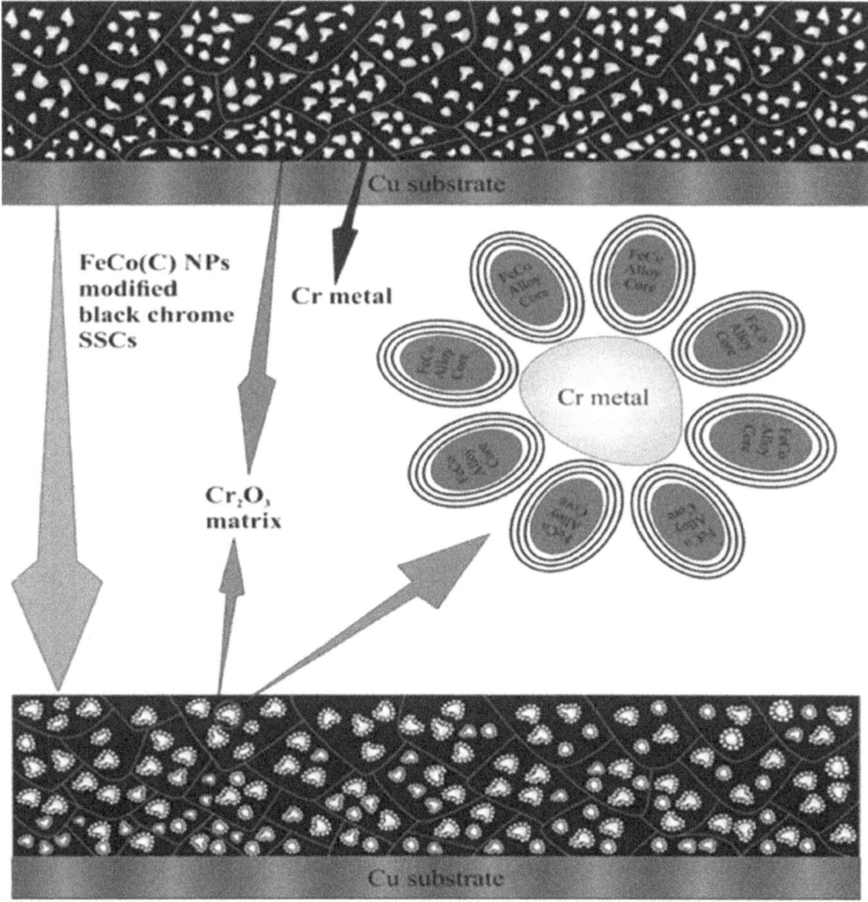

FIGURE 4.1 Schematic representation of black chrome and FeCo(C) NPs modified black chrome composite spectrally selective absorbers on Cu substrate.

4.3 EXPERIMENTAL PROCEDURE

Copper (Cu) and stainless-steel (SS) sheets (0.5 mm thick) were cut into 2.5 cm × 1 cm pieces and used as substrates for black chrome electrodeposition. All these substrates were mechanically polished with (number: 120) sand paper, followed by rinsing in distilled water. After polishing, substrates were degreased for approximately 10 min with trichloroethylene and acetone, followed by distilled water rinsing, and finally dried in air at room temperature. An infrared reflector layer of Ni had been deposited using a two-electrode electrochemical cell, where Copper (Cu), Ni/Cu (Ni-coated copper), and stainless-steel (SS) substrates were used as cathode, and PbSb alloy, with 2–5% Sb, was used as an anode. The electrolytic bath for Ni-coating consists of nickel sulphate (250 g/l), nickel chloride (60 g/l), boric acid (40 g/l), and pH of the solution has been maintained at 3.5 during the deposition process. The electrolytic bath was homogenised using magnetic stirrer at 200 rpm before initiating the electrochemical process under galvanostatic conditions. The current densities were varied from 0.2 A/cm^2 to 0.30 A/cm^2, and the time of deposition had also been varied from 60 s to 120 s to optimise the nickel layer on the copper substrate. The deposited Ni/Cu and Ni/SS samples were rinsed with deionised water to remove any residual debris and dried using hot air. The electrolytic bath for black chrome (BC) spectrally selective coating was composed of CrO_3 (275 g/l), NaF (0.2 g/l), and $NaNO_3$ (3 g/l) in water-based media. The electrolytic bath for FeCo(C) NPs modified black chrome spectrally selective coatings has been modified using 0.025, 0.05, and 0.10 wt.% of FeCo(C) NPs in BC electrolyte solution. The pH of the electrolytic bath has been maintained at 0.3 during all the deposition experiments.

BC and FeCo (C) NP-modified BC spectrally selective coatings are deposited using the two-electrode electrochemical cell as used for Ni layer deposition, at current densities ranging from 0.60 A/cm^2 to 0.80 A/cm^2 for 3,600 s. The deposited BC/Cu, BC/Ni/Cu, BC/SS, BC-FeCo (C)/Cu, BC-FeCo(C)/Ni/Cu, and BC-FeCo (C)/SS samples were rinsed with deionised water to remove any residual debris and dried using hot air.

We carried out X-ray diffraction (XRD) measurements to understand the development of phases and structures for the deposited thin-film cermet (metal–dielectric composite) coatings. XRD graphs were recorded in the range of 20° to 80° with 0.02° step size using copper K_α radiation (λ = 1.5406 Å). The microstructural and surface properties, such as roughness and grain size, were investigated using scanning electron microscope (SEM) and atomic force microscope (AFM) measurement systems. All AFM micrographs were recorded on 20 × 20 μm^2 and 10 × 10 μm^2 scanning areas, with 256 × 256 pixel resolutions. The samples were scanned in contact mode (CONTSCR cantilever tip with a radius of curvature < 10 nm) under ambient conditions. Elemental compositions were calculated using energy-dispersive X-ray (EDX) measurement system, equipped as an accessory with SEM equipment. Cary 4000 UV-Vis spectrophotometer was used to calculate absorptance of the selective absorber, and FTIR spectrophotometer was used to calculate the emittance values for the synthesised spectrally selective coatings. Electrochemical studies on these coatings were done in 3.5 wt.% NaCl (saline) solution employing Iviumstat spectro-electrochemical workstation to understand the corrosion properties of these coating structures. BC and BC-FeCo (C) NPs composite selective coatings deposited on Cu, Ni/Cu, and SS substrates act as the working electrode, Ag/AgCl and platinum act as reference and counter electrodes, respectively, for the corrosion and cyclic voltammetry studies. Tafel analysis has been used to calculate the corrosion current density, corrosion resistance, and corrosion rates. Thermal analyses of these coatings were carried out using Perkin–Elmer Simultaneous Thermal Analyser (STA) up to 900^{0}C temperatures under N_2 atmosphere at 10°C min^{-1} heating rate on pristine and corrosion-treated coating structures. These studies assisted in understanding the microstructural changes and their impact on solar thermal properties for developed spectrally selective coating structures.

4.4 RESULTS AND DISCUSSION

4.4.1 X-ray Diffraction Analysis

X-ray diffraction spectra are shown in Figure 4.2(A) a(i) and a(ii, iii, and iv) for pristine and FeCo (C) NPs modified black chrome composite spectrally selective coatings, deposited on copper substrates. Pristine black chrome (BC) selective coatings exhibit only hexagonal phase of chromium (Cr) at 2θ values of 38.40°, 43.72°, and 69.25°, corresponding to (100), (101), and (110) diffraction peaks (ICDD PDF #: 01–074–7045) (Figure 4.2 a(i)). However, with modifying FeCo (C) NPs, there is also a (110) diffraction peak at 2θ value of 44.40° of iron–cobalt–alloy phase (FeCo) (ICDD PDF #: 65–6829) (Figure 4.2 a(ii, iii, iv)) [Gupta *et al.*, 2014]. This suggests the presence of NPs, which modified the pristine black chrome selective coatings. Moreover, there are no diffraction peaks for chromium oxide (Cr_2O_3) crystallographic phase, suggesting that the chromium oxide is present in an amorphous form in these selective absorber coatings.

Figure 4.2 (B) shows the X-ray diffraction spectra of black chrome (Figure 4.2 b(i)) and modified black chrome (Figure 4.2 b(ii, iii, iv)) selective coatings on nickel-coated copper substrates. These XRD patterns also confirm the hexagonal chromium phase present in pristine black chrome (Figure 4.2 b(i)) with (110) and (200) diffraction peaks and iron–cobalt (FeCo) phase at 2θ values of 44.40° and 64.62° present with hexagonal chromium phase in modified black chrome selective coatings (Figure 4.2 b(ii, iii, iv)). The black chrome and modified black chrome coatings on SS substrate also show identical crystallographic phases like Cu and Ni/Cu substrates. The increased concentration of FeCo (C) nanoparticles in black chrome electrolyte resulted into an enhanced FeCo (C) concentration in black chrome thin films. This enhancement has been clearly observed in XRD graphs, where (110) plane FeCo alloy peaks became intense with increasing FeCo (C) concentration. In spite of large-fraction FeCo (C) nanoparticles, metallic chromium diffraction peak can be observed, suggesting that FeCo (C) NPs' covering may be very thin on metallic chromium, as indicated schematically in Figure 4.1.

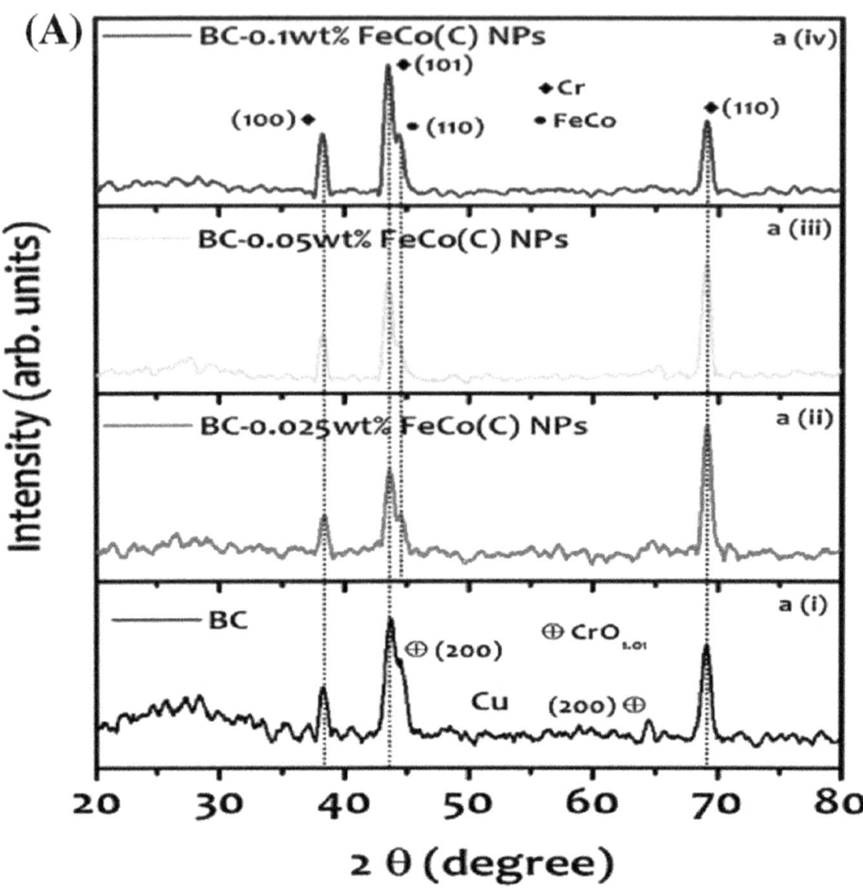

FIGURE 4.2(A) X-ray diffraction spectra of a(i) black chrome and a(ii, iii, iv) FeCo(C) NPs modified black chrome selective coatings with varaying wt.% of FeCo(C) NPs deposited on copper (Cu) substrate.

4.4.2 Microstructural Analysis

Scanning electron micrographs are shown in Figure 4.3 (a, b, c, and d) for black chrome and FeCo (C) NPs modified black chrome selective coatings. The black chrome coatings surface micrographs are smooth with micron-size grains, exhibiting large cracks, as can be seen in Figure 4.3a. The films became coarser with respect to pristine black chrome coatings after increasing FeCo (C) NP concentration in modified black chrome coatings. However, with enhanced coarseness, the films have shown an enhanced uniformity with lesser cracks and smaller grains. The thicknesses of these coatings are kept at higher side, approximately 10 μm, intentionally, as these films have been subjected to the thermal stresses and corrosion environments to understand their degradation and associated structure–property relation.

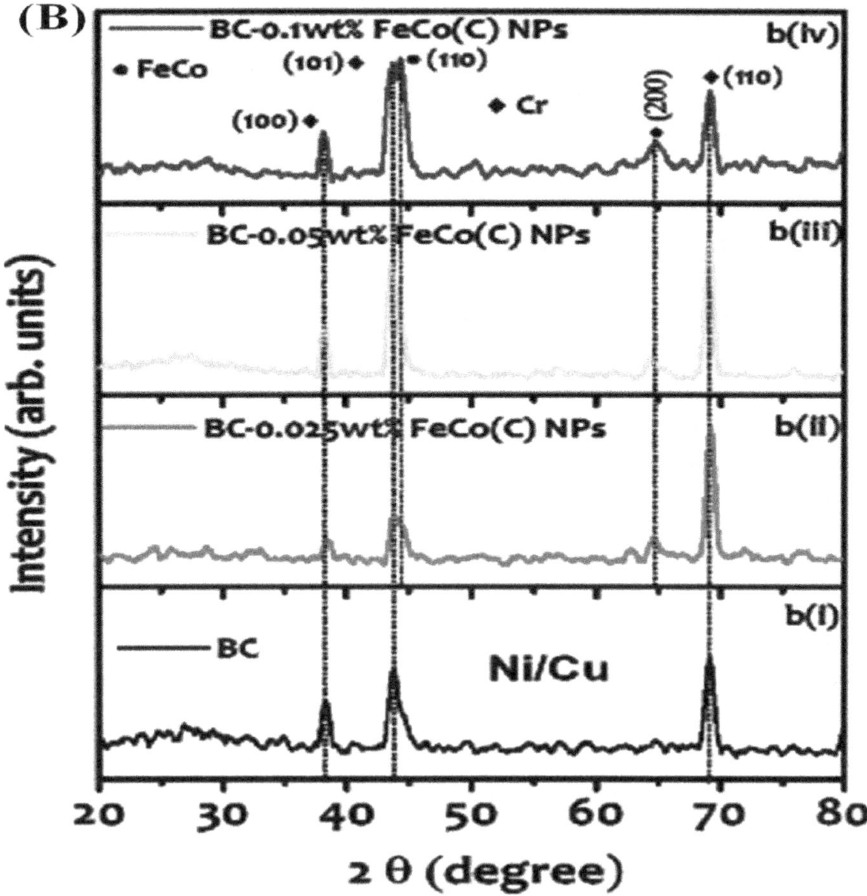

FIGURE 4.2(B) X-ray diffraction spectra of b(i) black chrome and b(ii, iii, iv) FeCo(C) NPs modified black chrome selective coatings with varaying wt.% of FeCo(C) NPs deposited on nickel coted copper (Ni/Cu) substrate.

Figure 4.4 shows the SEM micrographs of back chrome (a) and FeCo (C) NPs modified black chrome selective coatings (b, c, and d) on nickel-coated copper substrates. The surface micrographs of black chrome selective coatings suggest micron-size grains, similar to that of black chrome coatings on the copper substrates only. Here, also, cracks (Figure 4.4a) are visible throughout the film. The FeCo (C) NPs modified black chrome selective coatings show the coarser structure with less prominent cracks, as can be seen in Figure 4 (b, c, and d). The thicknesses of these coatings are approximately 12 μm, as shown in the cross-sectional SEM image for one of the structure, Figure 4.4 e.

In continuation, to further understand the microscopic structural details, AFM surface measurements are carried out for black chrome and FeCo (C) NPs modified black chrome selective coatings deposited on nickel-coated copper substrate.

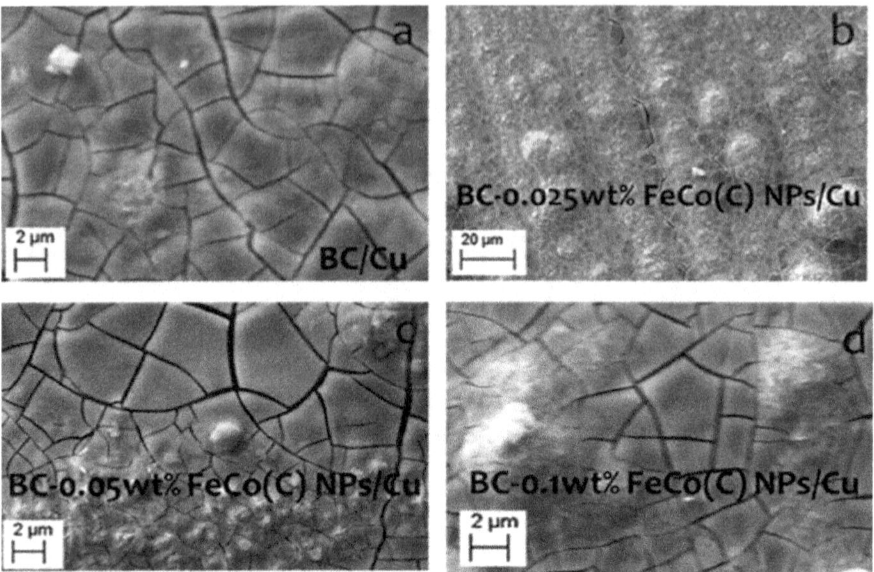

FIGURE 4.3 (a) SEM micrographs of black chrome and (b, c, and d) FeCo(C) NPs modified black chrome with varying wt.% of FeCo(C) NP-selective surfaces deposited on copper substrates.

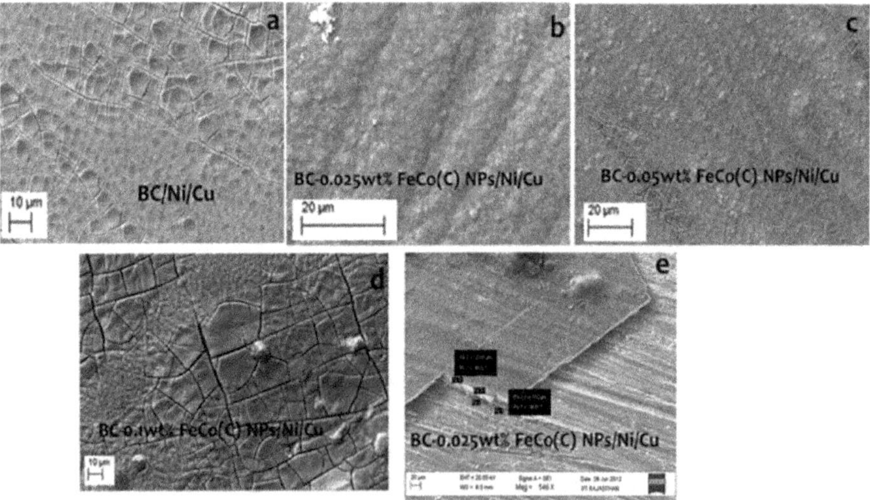

FIGURE 4.4 (a) SEM micrographs of black chrome and (b, c, and d) FeCo(C) NPs modified black chrome with varying wt.% of FeCo(C) NP-selective surfaces deposited on copper substrate. (e) SEM cross-section of thickness measurements.

These AFM images are shown in Figure 4.5 (A, B, C, and D). Black chrome selective coatings show striped wrinkle-like surface morphology (Figure 4.5a), whereas island-type surface morphology has been observed for FeCo (C) NPs modified black chrome selective coatings (Figure 4.5b). In addition, FeCo (C) NPs modified black chrome selective coatings exhibit less average surface roughness (R_a) as compared to the pristine black chrome selective coatings (Figure 4.5a), consistent with SEM observations, as discussed earlier.

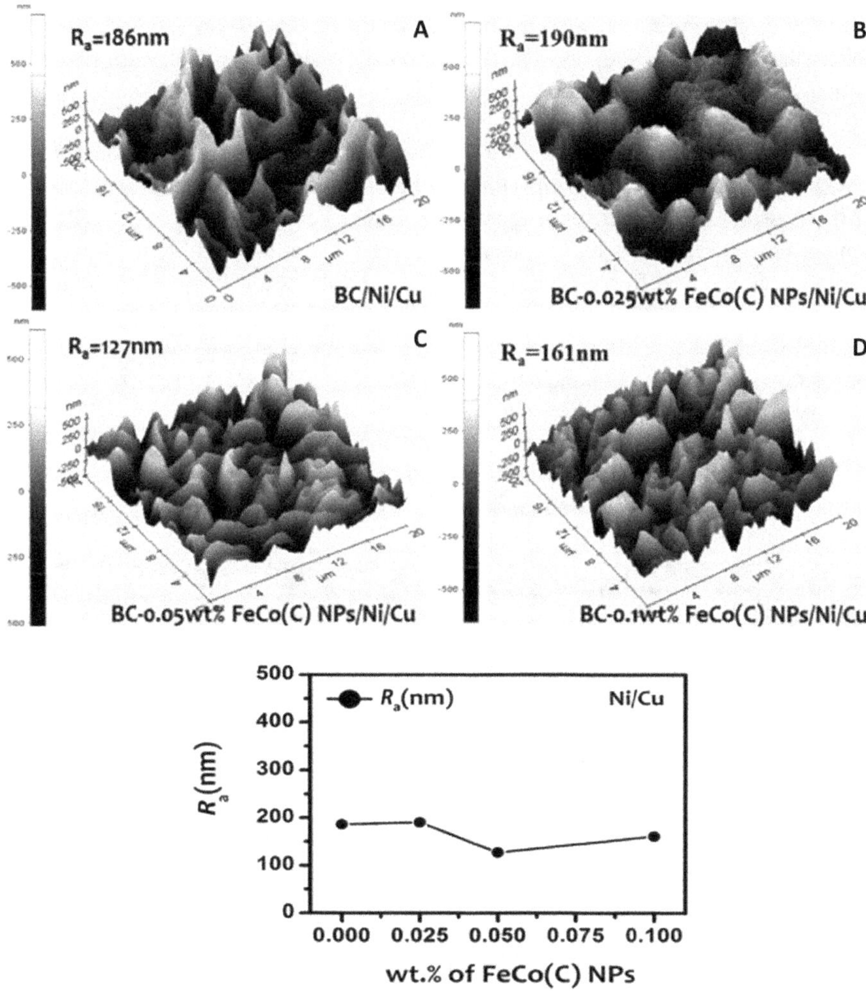

FIGURE 4.5 (a) AFM surface morphology (A) of black chrome and (B, C, and D) FeCo(C) NPs modified black chrome with varying wt.% of FeCo(C) NP-selective surfaces deposited on nickel-coated copper substrate. (b) Average roughness versus wt.% of FeCo(C) NPs.

4.4.3 Elemental Analysis

The elemental results are summarised in Figure 4.6 (a, b, c, and d) for black chrome and different wt.% of FeCo (C) NPs modified black chrome selective coatings on copper substrates. These energy-dispersive X-ray spectra of black chrome selective coating show the chromium (Cr) and oxygen (O) elemental peaks only (Figure 4.6a). The corresponding elemental compositions are listed in the insets of respective spectra in both weight% and atomic% (Figure 4.6a). The observed elemental fractions of oxygen and chromium are 74.58 atm.%, and 25.42 atm.%, respectively, suggesting the presence of chromium oxide phase in these coatings. The FeCo (C) NPs modified black chrome selective coatings show additional Fe elemental peak with Cr and O peaks, similar to that of pristine black chrome coating (Figure 4.6b, c, and d). The 0.025 wt.% FeCo(C) NPs modified black chrome coatings have shown very poor Fe, Co, and C elemental fractions of approximately 0.05 atm.%, 0.06 atm.%, and 8.04 atm.% respectively (Figure 4.6b). This may be due to the small fraction of FeCo (C) in the coatings. Moreover, with increasing weight percent of FeCo (C) NPs, iron content has increased, as can be seen in Figure 4.6(c and d). These measurements were carried out at different sections across the surfaces for different coatings, and it was observed that even with the least wt.% FeCo (C) NPs modified black chrome spectrally selective coatings, Fe, Co, and C elemental compositions are distributed uniformly, suggesting a homogeneous distribution of FeCo (C) nanoparticles within the coating structures. The elemental composition of black chrome and 900^{0}C (C) NPs modified black chrome selective coatings on nickel-coated copper substrates has also shown identical elemental distribution and results as discussed by Usmani *et al.* in their published work [Usmani and Harinipriya, 2013].

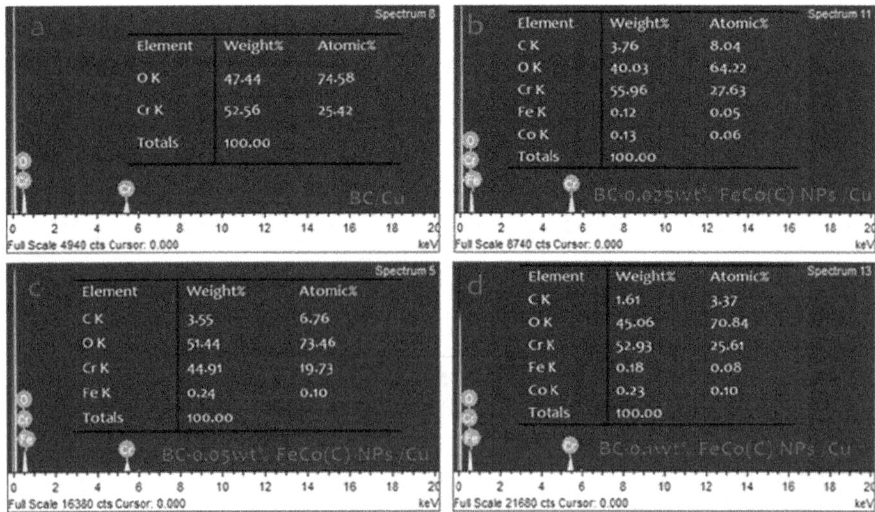

FIGURE 4.6 EDS elemental analysis of (a) black chrome and (b, c, and d) FeCo(C) NPs modified black chrome with varying wt.% of FeCo(C) NP-selective surfaces deposited on the copper substrate.

4.4.4 Optical Properties

UV-Vis reflectance spectra were collected in 200–800 nm wavelength range and are plotted in Figure 4.7 (a and b) for both black chrome and FeCo (C) NPs modified black chrome selective coatings on copper and nickel-coated copper substrates. These reflectance data are used to calculate the absorptance values and are summarised in the insets of respective reflectance plots against the weight percent of

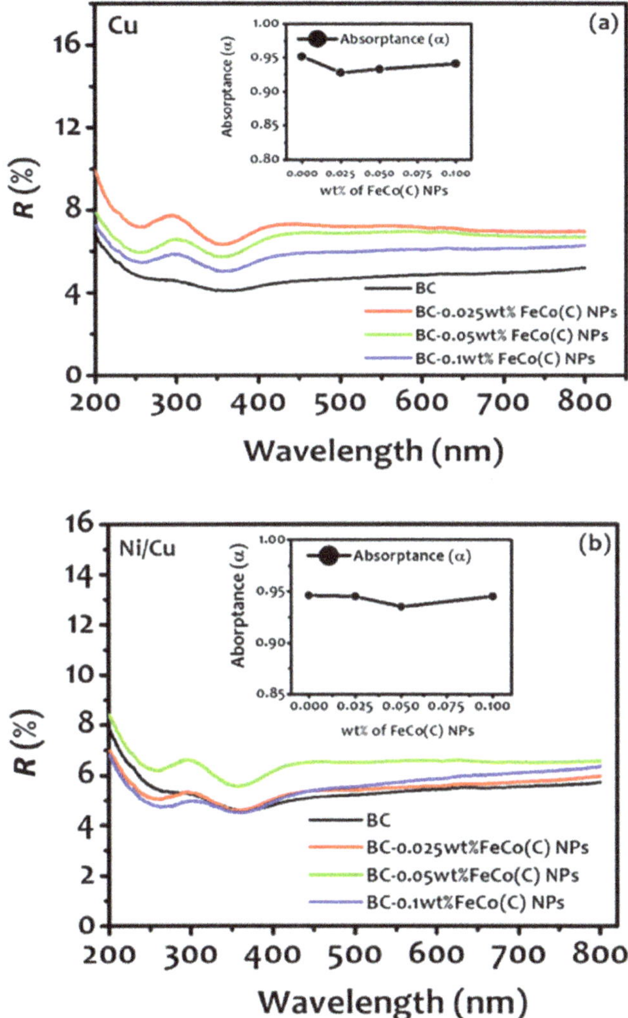

FIGURE 4.7 Plot of reflectance spectra versus wavelength of black chrome and FeCo(C) NPs modified black chrome with varying wt.% of FeCo(C) NP-selective coatings deposited on (a) Cu and (b) nickel-coated Cu substrates. Calculated absorptance values versus varying wt.% of FeCo(C) NPs are represented in the inset in the graphs.

FeCo (C) NPs coating (Figure 4.7 (a and b) insets) on both Cu and nickel-coated Cu substrates. Absorptance values are 0.952, 0.928, 0.933, and 0.941 of BC and FeCo (C) NP-modified BC on Cu substrates and are 0.946, 0.945, 0.935, and 0.945 of BC and Cr_2O_3 (C) NP-modified BC on nickel-coated copper substrates. These absorptance values are relatively insensitive to FeCo (C) modifications, suggesting that the addition of FeCo (C) metallic nanoparticles has no significant contribution to the absorption as these values are close to the pristine black chrome structures. The observed broad minima at approximately 250 nm and at approximately 350 nm correspond to intraband 3d transitions in Cr metal and FeCo alloys and are consistent with reported literature [Quinten, 2011]. These minima suggest that along with Cr, FeCo nanoparticles may also be contributing to the strong absorption more than approximately 0.95 in the solar spectrum region.

The emittance values of these coating structures are calculated using infrared reflectance and are plotted in Figure 4.8, for BC and FeCo (C) NP-modified BC in 2.5–25 μm wavelength range. The calculated emittance values are presented in the inset of Figure 4.8. These emittance values of black chrome and FeCo (C) NPs modified black chrome selective coatings are relatively higher and show variation from 0.47 to 0.66. The large thickness approximately 10 μm of these structures is the main reason for such high emittance values. It can be decreased by up to 0.1 or close to the literature values for black chrome structure by reducing the thickness and optimising the metallic content. However, the optical properties are not the main objective for this work, and that's why thicknesses of these coatings were kept large intentionally to understand the thermal stability of these coatings through thermogravimetric analysis and the microscopic origin of degradation using different structure–property correlations.

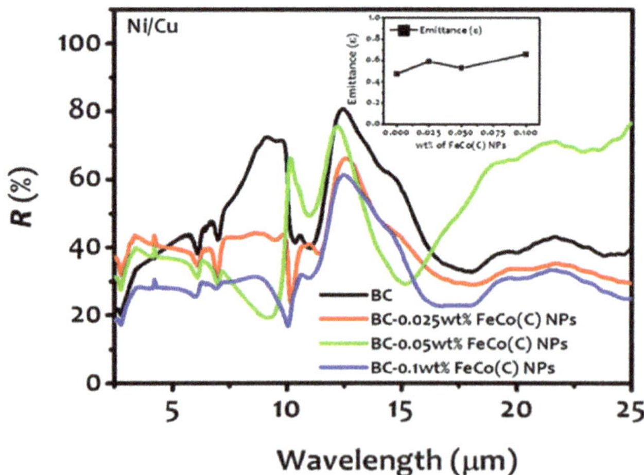

FIGURE 4.8 Plot of reflectance spectra versus wavelength in the range of 2.5–25 μm of black chrome and of FeCo(C) NPs modified black chrome with varying wt.% of FeCo(C) NP-selective coatings deposited on nickel-coated Cu substrates. Calculated emittance values versus varying wt.% of FeCo(C) NPs have been represented in the inset in the graphs.

4.4.5 Electrochemical Measurements

Environmental degradation, especially due to the humidity, salinity, and temperature, etc., may hamper the response of black chrome spectrally selective coatings. Thus, it is important to understand the degradation mechanism, which may assist in developing environmentally stable coating structures. Corrosion studies have been carried out using linear voltammetry measurements, and the related analysis is discussed next.

4.4.5.1 Cyclic Voltammetry

The electrochemical studies, such as cyclic voltammetry (CV), were carried out in 3.5 wt.% NaCl saline solution to understand the redox process and its effect on the selective coatings. The cyclic voltammograms are shown in Figure 4.10, for black chrome and different wt.% FeCo (C) NPs modified black chrome coatings on the copper substrate. These measurements are recorded at different scan rates (from 5 to 200 mVs^{-1}) in 3.5 wt.% NaCl (saline) solution.

These voltammograms (Figure 4.9 (i, ii, iii, and iv)) suggest that the current gap between anodic and cathodic scan increases with increasing scan rates. This could be attributed to the change in the diffusion rates of ions or mass transfer from the bulk of the solution to the electric double layer (the interface between an electrolyte solution and an electrode). A schematic model of electrode–solution

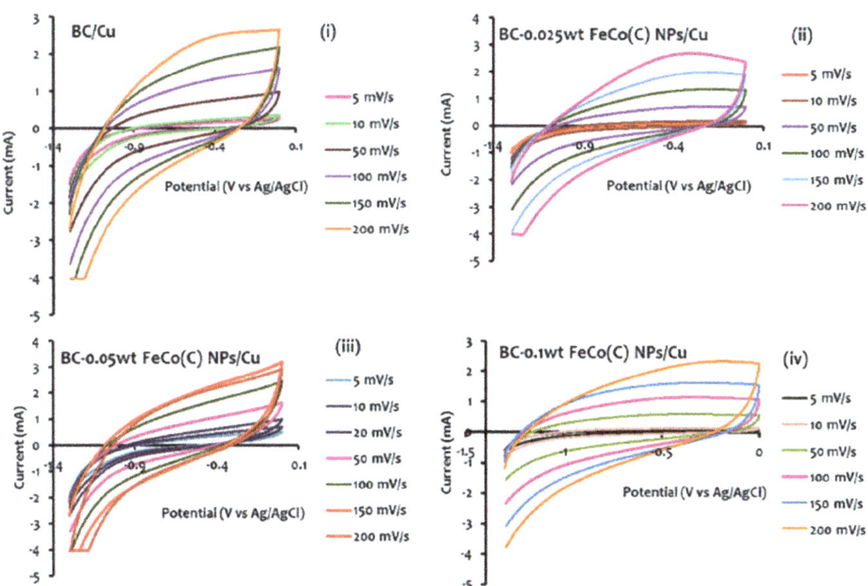

FIGURE 4.9 Cyclic voltammogram of (i) black chrome and (ii, iii, and iv) FeCo (C) NPs modified black chrome selective coatings with varying wt.% of FeCo(C) NPs deposited on copper substrate, recorded with different scan rates (from 5 to 200 mVs^{-1}), in 3.5 wt.% NaCl (saline) solution.

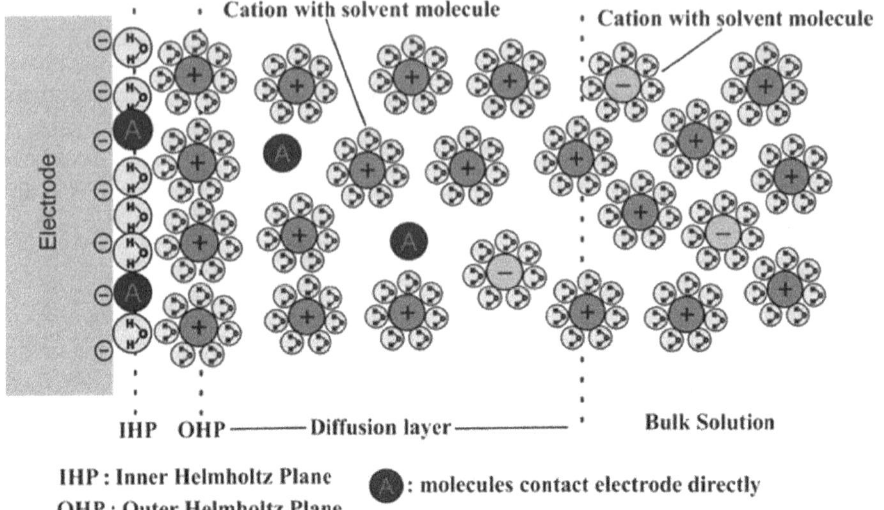

FIGURE 4.10 Schematic of electrode–solution interphase (electric double layer).

Source: Kissinger and Heineman (1996b)

interface (double layer) is illustrated in Figure 4.10. This explains the contribution of bulk solution in the formation of inner and outer Helmholtz plane, the diffusion layer in contact with electrode during cyclic voltammetry measurements. More surprisingly, these cyclic voltammetry measurements do not exhibit specific redox peaks corresponding to any electrochemical oxidation/reduction process, indicating that the coatings are electrochemically stable even in highly saline environmental conditions.

Figure 4.11 shows the cyclic voltammograms of black chrome and FeCo (C) NPs modified black chrome with varying wt.% of FeCo (C) NPs selective coatings on nickel-coated copper substrate, recorded at different scan rates (from 5 to 200 mVs^{-1}). These voltammograms also do not exhibit any specific redox peak corresponding to any electrochemical oxidation or reduction, suggesting that these coatings are also electrochemically stable in the highly saline environment.

4.4.5.2 Linear Sweep Voltammetry: Corrosion

Tafel studies are carried out by employing linear sweep voltammetry of black chrome and FeCo (C) NPs modified black chrome selective coatings deposited on copper and nickel-coated copper substrates as cathode, Pt coil as counter electrode, and Ag/AgCl as reference electrode in highly saline/marine environment of 3.5 wt.% NaCl electrolyte solution. The scan rate was varied from 1 mVs^{-1} to 100 mVs^{-1} to understand the corrosion impact on solar selective coatings in saline environments and the effect of nanoparticles on corrosion on black chrome coatings. The potentiodynamic polarisation curves of BC and FeCo (C) NPs modified black chrome selective coatings on copper substrates are shown in Figure 4.12 (left panel (i, ii,

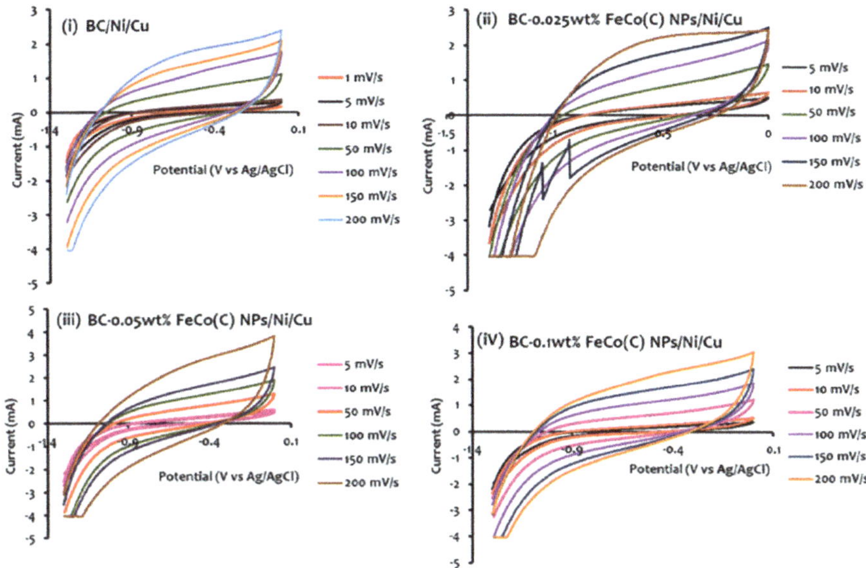

FIGURE 4.11 Cyclic voltammogram of (i) black chrome and (ii, iii, iv) FeCo (C) NPs modified black chrome selective coatings with varying wt.% of FeCo(C) NPs deposited on nickel coated copper substrate, recorded with different scan rates (from 5 to 200 mVs^{-1}), in 3.5 wt.% NaCl (saline) solution.

iii, and iv)) and used to calculate the numerous corrosion parameters. The corrosion potential (E_{corr}), corrosion current density (i_{corr}), corrosion resistance (R_P), and corrosion rate (C. Rate) are summarised in respective tables of Figure 4.12 (right panel (i, ii, iii, and iv)). The corrosion potential becomes more negative (−0.8467 V to −1.1270 V) with increasing the scan rate (from 1 mVs^{-1} to 100 mVs^{-1}) for the pristine black chrome coatings (Figure 4.11 (left panel)). The corrosion current density increases from 9.119 to 389.0 μA/cm^2, with an increasing scan rate from 1 mVs^{-1} to 100 mVs^{-1}. Similarly, corrosion rate increases from 0.2598 to 11.08 mm/y, whereas corrosion resistance gets reduced from 7.805 to 0.0668 kΩ, with the scan rate increasing from 1 mVs^{-1} to 100 mVs^{-1}.

The potentiodynamic curves are shown in Figure 4.12(ii) for 0.025 wt.% FeCo (C) NPs modified black chrome selective coating, and respective corrosion parameters are summarised in Table (ii). Corrosion resistance value is relatively lower for 0.025 wt.% modified black chrome selective coatings as compared to pristine black chrome at low scan rate (1–5 mVs^{-1}), while at higher scan rate (20 mVs^{-1} to 100 mVs^{-1}), the corrosion resistance values become larger (0.5667 kΩ to 0.204 kΩ) as compared to that of pristine black chrome selective coatings (0.5082 to 0.06684 kΩ). These measurements suggest that 0.025 wt.% FeCo (C) NPs modified black chrome coatings exhibit an enhanced corrosion resistance as compared to that of pristine black chrome structures. Similar behaviour has been observed for 0.05 wt.% and 0.10 wt.% FeCo (C) NPs modified black chrome spectrally selective coatings on Cu substrates,

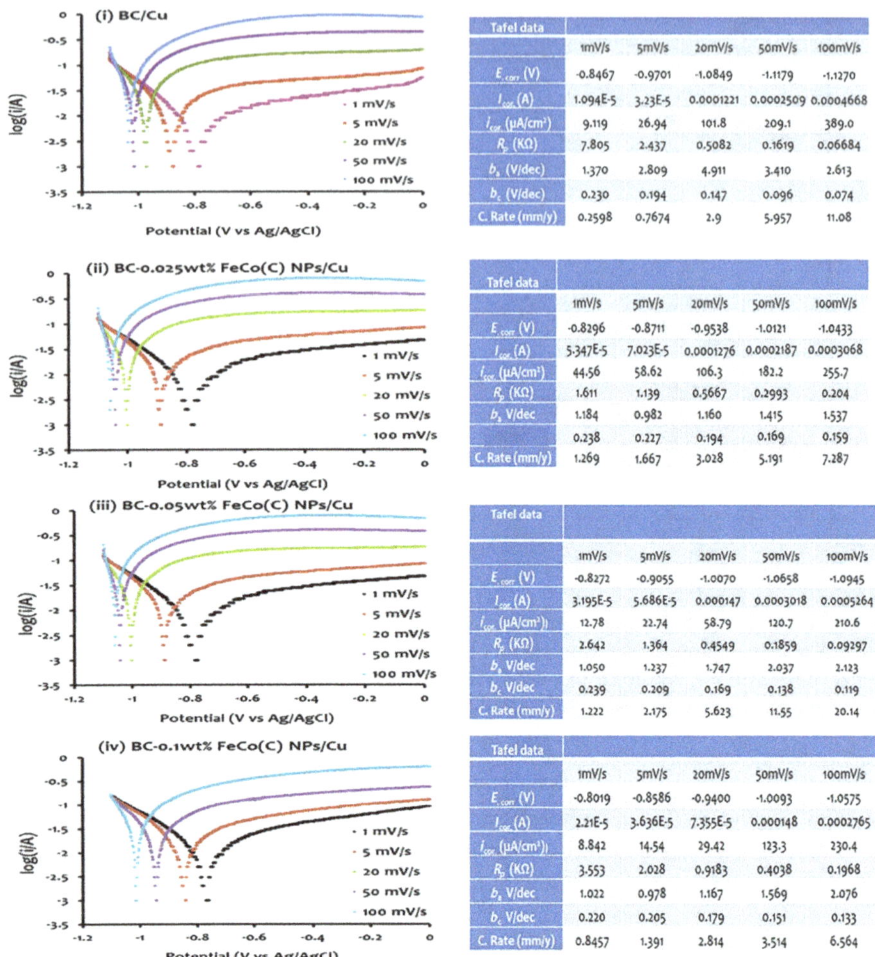

Tafel data					
	1mV/s	5mV/s	20mV/s	50mV/s	100mV/s
E_{corr} (V)	-0.8467	-0.9701	-1.0849	-1.1179	-1.1270
$I_{cor.}$ (A)	1.094E-5	3.23E-5	0.0001221	0.0002509	0.0004668
$i_{cor.}$ (μA/cm²)	9.119	26.94	101.8	209.1	389.0
R_p (KΩ)	7.805	2.437	0.5082	0.1619	0.06684
b_a (V/dec)	1.370	2.809	4.911	3.410	2.613
b_c (V/dec)	0.230	0.194	0.147	0.096	0.074
C. Rate (mm/y)	0.2598	0.7674	2.9	5.957	11.08

Tafel data					
	1mV/s	5mV/s	20mV/s	50mV/s	100mV/s
E_{corr} (V)	-0.8296	-0.8711	-0.9538	-1.0121	-1.0433
$I_{cor.}$ (A)	5.347E-5	7.023E-5	0.0001276	0.0002187	0.0003068
$i_{cor.}$ (μA/cm²)	44.56	58.62	106.3	182.2	255.7
R_p (KΩ)	1.611	1.139	0.5667	0.2993	0.204
b_a V/dec	1.184	0.982	1.160	1.415	1.537
	0.238	0.227	0.194	0.169	0.159
C. Rate (mm/y)	1.269	1.667	3.028	5.191	7.287

Tafel data					
	1mV/s	5mV/s	20mV/s	50mV/s	100mV/s
E_{corr} (V)	-0.8272	-0.9055	-1.0070	-1.0658	-1.0945
$I_{cor.}$ (A)	3.195E-5	5.686E-5	0.000147	0.0003018	0.0005264
$i_{cor.}$ (μA/cm²))	12.78	22.74	58.79	120.7	210.6
R_p (KΩ)	2.642	1.364	0.4549	0.1859	0.09297
b_a V/dec	1.050	1.237	1.747	2.037	2.123
b_c V/dec	0.239	0.209	0.169	0.138	0.119
C. Rate (mm/y)	1.222	2.175	5.623	11.55	20.14

Tafel data					
	1mV/s	5mV/s	20mV/s	50mV/s	100mV/s
E_{corr} (V)	-0.8019	-0.8586	-0.9400	-1.0093	-1.0575
$I_{cor.}$ (A)	2.21E-5	3.636E-5	7.355E-5	0.000148	0.0002765
$i_{cor.}$ (μA/cm²))	8.842	14.54	29.42	123.3	230.4
R_p (KΩ)	3.553	2.028	0.9183	0.4038	0.1968
b_a V/dec	1.022	0.978	1.167	1.569	2.076
b_c V/dec	0.220	0.205	0.179	0.151	0.133
C. Rate (mm/y)	0.8457	1.391	2.814	3.514	6.564

FIGURE 4.12 Left panel: potentiodynamic polarisation curve of (i) black chrome and (ii, iii, and iv) FeCo(C) NPs modified black chrome solar selective coatings with varying wt.% of FeCo(C) NPs deposited on copper substrates in 3.5 wt.% (saline) NaCl solution, with varying scan rates (1 mVs^{-1} to 100 mVs^{-1}). Right panel (Tables) represents the estimated corrosion parameter of black chrome and FeCo(C) NPs modified black chrome selective coatings with varying wt.% of FeCo(C) NPs on copper substrates through Tafel fitting of polarisation curves.

as shown in Figure 4.13, where corrosion resistance has been plotted against the scan rates. These results suggest that corrosion rate has reduced to half (6.564 mm/y) for 0.1 wt.% FeCo (C) NPs modified black chrome selective coatings, as compared to the pristine black chrome (11.08 mm/y) at 100 mVs^{-1} scan rate.

The observed enhancement in corrosion resistance for FeCo (C) NPs modified black chrome structures is consistent with the measured lower corrosion rate for FeCo (C) NPs modified black chrome. This also suggests that FeCo (C) NPs may be providing a protective coating on chromium metallic structures in black chrome

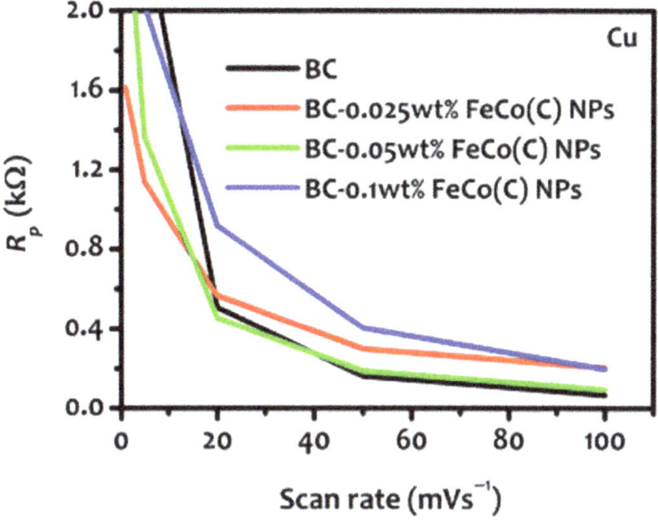

FIGURE 4.13 Corrosion resistance versus scan rate of black chrome and FeCo(C) NPs modified black chrome with varying wt.% of NPs selective coatings on copper substrates.

coatings and thus avoiding direct access to the corrosive media. These observations corroborate the proposed hypothesis, as shown in Figure 4.1. This enhancement in corrosion resistance may be due to the environmentally stable graphite-encapsulated FeCo (C) NPs and their effective distribution in the cermet matrix. Corrosion study on black chrome and FeCo (C) NPs modified black chrome selective coatings on nickel-coated copper substrates also showed a similar behaviour as on copper substrates, and details are summarised in Usmani *et al.* reference [Usmani and Harinipriya, 2013].

4.4.6 Thermal Stability

Thermal stability tests for these coating structures are carried out using thermogravimetric analysis in inert environmental conditions. The measurements are summarised in Figure 4.14 in the form of percent weight change as a function of temperature for black chrome and FeCo (C) NPs modified black chrome selective coatings, deposited on copper and nickel-coated copper substrates, respectively. Pristine black chrome coatings on copper substrate showed an initial weight change approximately 0.85% up to 350°C and remained nearly constant up to 700°C in N_2 environment, as can be seen in Figure 4.14 (a). The initial percentage weight change is attributed to the loss of volatile components such as unburnt organics during the synthesis and post-synthesis processes. However, the weight gain observed after 700°C is attributed to the formation of higher molecular weight compounds such as chromium nitride (CrN), nitrous oxides (N_2O), or even Cr_2O_3 due to the oxidation of metallic chromium in black chrome matrix. The instability of nitrogen molecule at a higher temperature may be responsible for the formation of nitride compounds in nitrogen environmental conditions, and residual oxygen may lead to the oxidation of metallic chromium. FeCo(C) NPs modified black chrome selective coatings on copper substrate showed an enhanced thermal stability as compared to the pristine one, as can be seen in Figure 4.14 (a). A relatively lower weight change of approximately 0.31% has been observed initially, from 120°C to 220°C with FeCo(C) NPs modified black chrome selective coatings, and is attributed to the loss of volatile components,

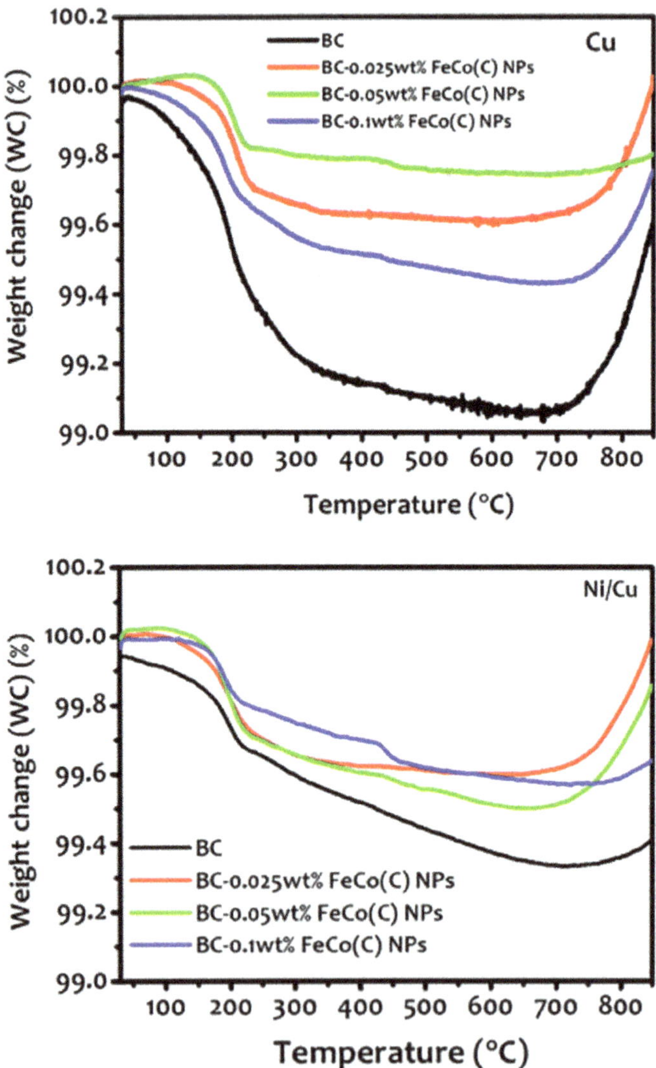

FIGURE 4.14 Percentage weight change as a function of temperature for black chrome and FeCo(C) NPs modified black chrome selective coatings with varying wt.% of NPs deposited on (a) copper and (b) nickel-coated copper substrate, in an inert atmosphere.

as also observed in pristine BC structure. The coating is further relatively stable up to 730°C. Here also, the observed weight gain after 730°C is attained is due to the formation of degradation products of higher molecular weight, as in the case of pristine black chrome coatings. Higher thermal stability of FeCo (C) NPs modified black chrome selective coatings is attributed to the protective coating of thermally stable FeCo (C) NPs on metallic chromium in black chrome matrix. These results indicate that 0.05 wt.% FeCo (C) NPs modified black chrome selective coatings are relatively better thermally stable as compared to 0.025 and 0.1 wt.% of nanoparticles modified black chrome selective coatings.

Figure 4.14 (b) shows the percentage weight change of black chrome and FeCo (C) NPs modified black chrome selective coatings on nickel-coated copper substrates, as a function of temperature. These measurements also suggest that the thermal stability of FeCo (C) NPs modified black chrome selective coatings on nickel-coated copper substrates is better as compared to that of black chrome structures only. More surprisingly, spectrally selective structures on Ni/Cu show an enhanced thermal stability as compared to that of bare Cu substrate. The relative thermal stability of Ni in Ni/Cu provides better thermally stable spectrally selective structures as compared to that of Cu substrates only. Pristine black chrome coatings on nickel coated copper substrate showed an initial weight loss like coatings on an uncoated copper substrate. The observed initial percentage weight change is attributed to the loss of volatile components such as unburnt organics, as observed and discussed previously for coatings on Cu substrates. Here also, the observed weight gain after 800°C is attained is attributed to the formation of higher molecular weight compounds such as chromium nitride (CrN), nitrous oxides (N_2O), or even Cr_2O_3 due to the instability of nitrogen molecule at higher temperatures and oxidation of metallic chromium in black chrome matrix. Moreover, graphite-encapsulated FeCo-modified black chrome selective films showed a relatively better thermal stability, Figure 4.14 (b), similar to that of uncoated nickel copper substrate.

4.5 ANALYSIS AFTER ELECTROCHEMICAL MEASUREMENTS

The structural, microstructural, and compositional analysis of this corrosion-treated black chrome and FeCo(C) NPs modified black chrome spectrally selective coatings have been done in detail to understand the possible degradation and structure–property correlation, as discussed next.

4.5.1 Structural, Microstructural, and Elemental Analysis after Electrochemical Measurements

X-ray diffraction spectra of cyclic voltammetry-treated black chrome and FeCo (C) NPs modified black chrome selective coatings on copper substrates are shown in Figure 4.15 a(i, ii, iii, and iv). These XRD measurements suggest that there are no structural changes in case of FeCo (C) NPs modified black chrome coatings, and diffraction spectra are nearly identical to that of the pristine ones, as shown and discussed in Section 4.4.1. The pristine black chrome structures exhibited additional diffraction peaks at 2θ values of 44.34° (110) and 64.80° (200), which

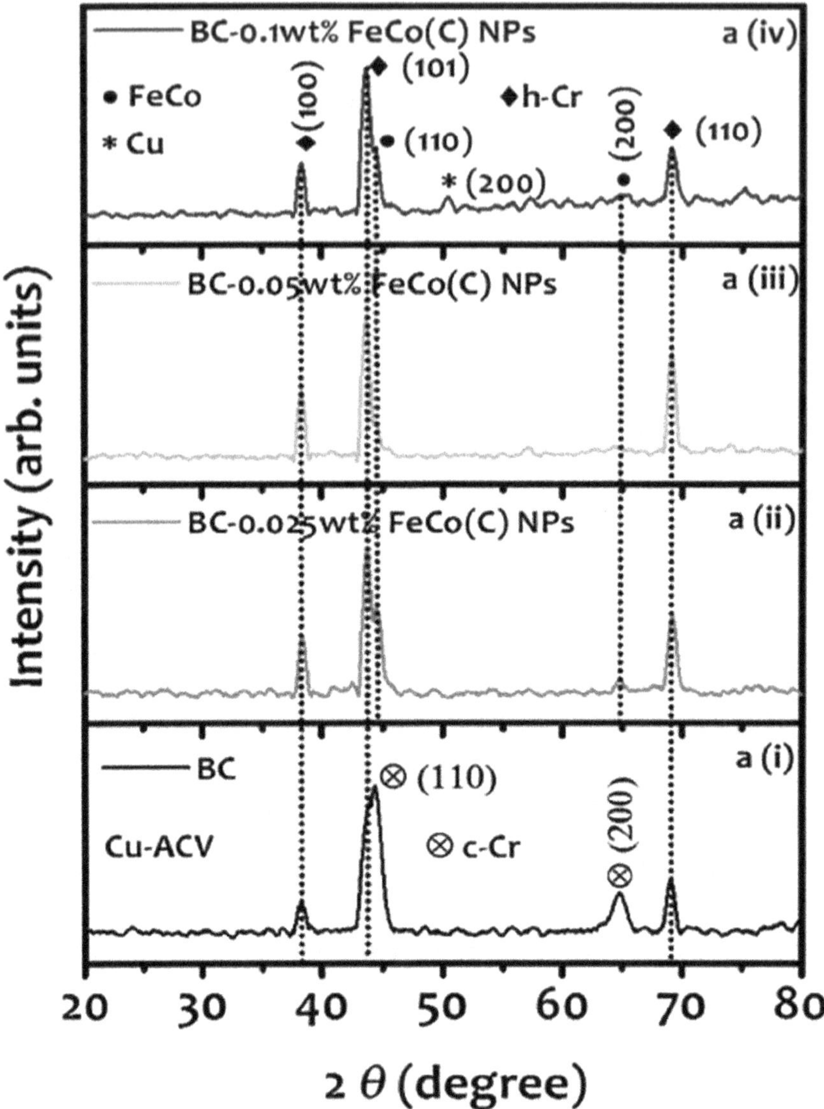

FIGURE 4.15 X-ray diffraction spectra of (a and b) (i) black chrome and (a and b) (ii, iii, iv) FeCo(C) NPs modified black chrome selective coatings with varying wt.% of FeCo(C) NPs deposited on copper substrates (a) after cyclic voltammetry measurements and (b) after corrosion measurements.

correspond to the cubic chromium phase (ICDD #: 01–077–7591) (Figure 4.15 a(i)), after corrosion treatments. The onset of additional chromium diffraction peaks in pristine black chrome suggests that the corrosion has initiated and degraded black chrome layer, whereas corrosion has the least impact on FeCo(C) NPs modified black chrome coating structures.

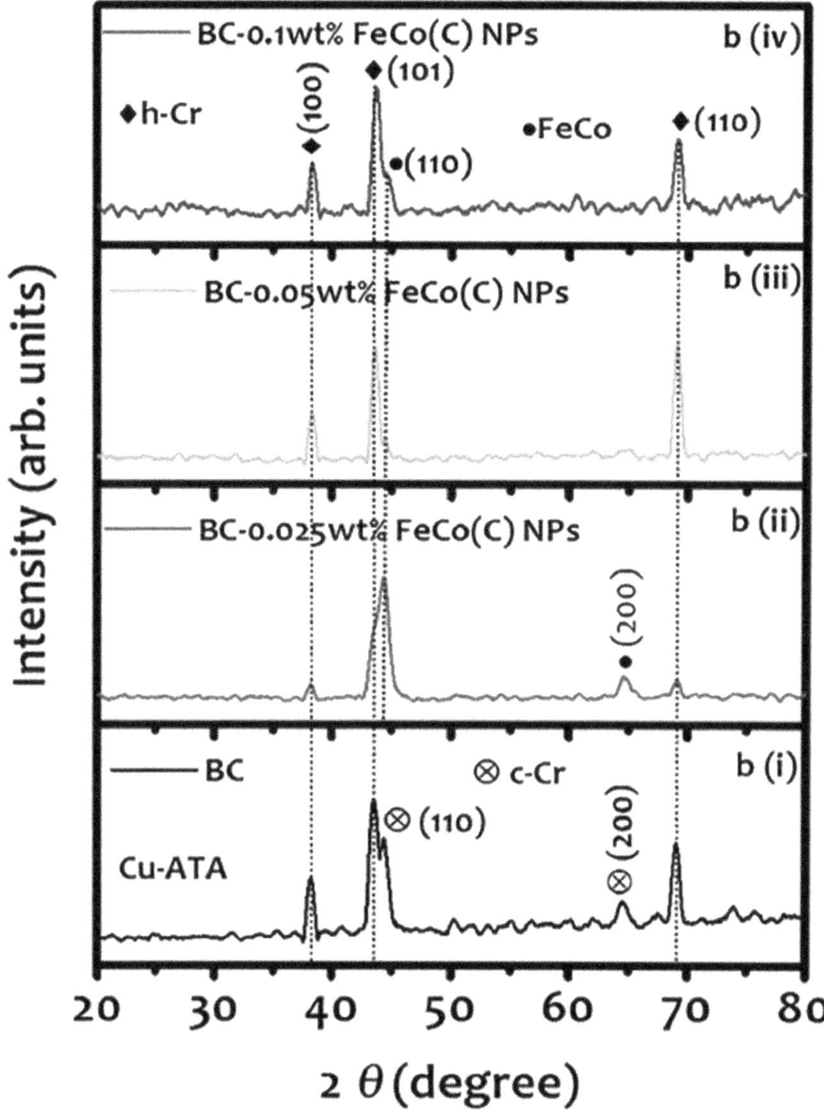

FIGURE 4.15 (Continued)

Figure 4.15 b(i, ii, iii, and iv) shows the X-ray diffraction spectra of black chrome and FeCo (C) NPs modified black chrome selective coatings deposited on copper substrates, after corrosion measurements. Similar XRD patterns have been recorded after the corrosion measurements for FeCo(C) NPs modified black chrome structures, suggesting relative inertness to the corrosion as compared to pristine black chrome, where additional chromium diffraction peaks have been recorded, as shown in Figure 4.15 b(i). Structural properties exhibit similar crystallographic phases for black chrome and FeCo(C) NPs modified black chrome selective coatings on nickel-coated copper and SS substrates as on copper substrates.

Figure 4.16 shows the SEM surface microstructure of black chrome (Figure 4.16 (a)(i)) and FeCo (C) NPs modified black chrome selective coatings on copper substrate (Figure 4.16 a(ii, iii, and iv)), after cyclic voltammetry measurements. These measurements suggest that the grain size of pristine black chrome coating has changed from micro to needle-shaped nano grains after cyclic voltammetry measurements (Figure 4.16 (a)(i)). Moreover, cracks on pristine black chrome selective coatings are completely filled and became much denser, after cyclic voltammetry measurements. FeCo (C) NPs modified black chrome selective coatings (0.025 wt.%) surface microstructure, Figure 4.16 a(ii), became much smoother, and the cracks were partially filled with composite particles after cyclic voltammetry measurements. Surface microstructure of 0.05 wt.% FeCo (C) NPs modified black chrome selective coatings, Figure 4.16 a(iii), is showing the hybrid nature containing partially smooth and partially denser nano grains. The cracks were filled with denser nanoparticles, whereas the already existing islands at smooth regions of the coatings also co-existed (Figure 4.3 (c)). Figure 4.16 a(iv) shows the surface microstructure of 0.1 wt.% FeCo (C) NPs modified black chrome selective coatings, after cyclic voltammetry measurements. The cracks are filled, and the grains have changed drastically from micro to needle-shaped nano grains. This anomalous behaviour may be attributed to the variation in the diffusion of ions and mass transfer rates of the active species in the spectrally selective structures.

Figure 4.16 b(i) shows the surface microstructure of pristine black chrome selective coatings on a copper substrate, after corrosion measurements. This SEM study reveals that the particle size changed drastically from micron size grains to nano-sized grains, and the geometry of the surface becomes more needle like, indicating the corrosion impact on these surfaces. However, the surface microstructure of 0.025 wt.% modified black chrome selective coatings (Figure 4.16 b(ii)) suggests that cracks on the surface are almost covered, with enhanced surface roughness. FeCo (C) NPs (0.05 wt.%) modified black chrome selective coatings surface microstructure, Figure 4.16 b(iii), suggests the entire coverage of cracks and boils due to the corrosion effects on the surface. Surface microstructure of 0.1 wt.% FeCo (C) NPs modified black chrome selective coatings, Figure 4.16 b(iv), does not show any effect on it, suggesting that corrosion has the least impact on this structure. From the earlier SEM microstructural analysis, it is evident that the stability of FeCo (C) NPs modified black chrome coatings has enhanced even in saline environments. Similar analysis of these coatings on Ni/Cu substrates is discussed in detail by Usmani *et al.* [Usmani and Harinipriya, 2013].

After cyclic voltammetry measurements, atomic force microscopy (AFM) measurements are shown in Figure 4.17 a(i, ii, iii, and iv) for black chrome and FeCo (C) NPs modified black chrome selective coatings on the copper substrate. These measurements also confirmed similar surface microstructural features, as observed in SEM micrographs. Average surface roughness has increased for 0.025 and 0.05 wt.% FeCo (C) NPs modified black chrome selective coatings, as shown graphically

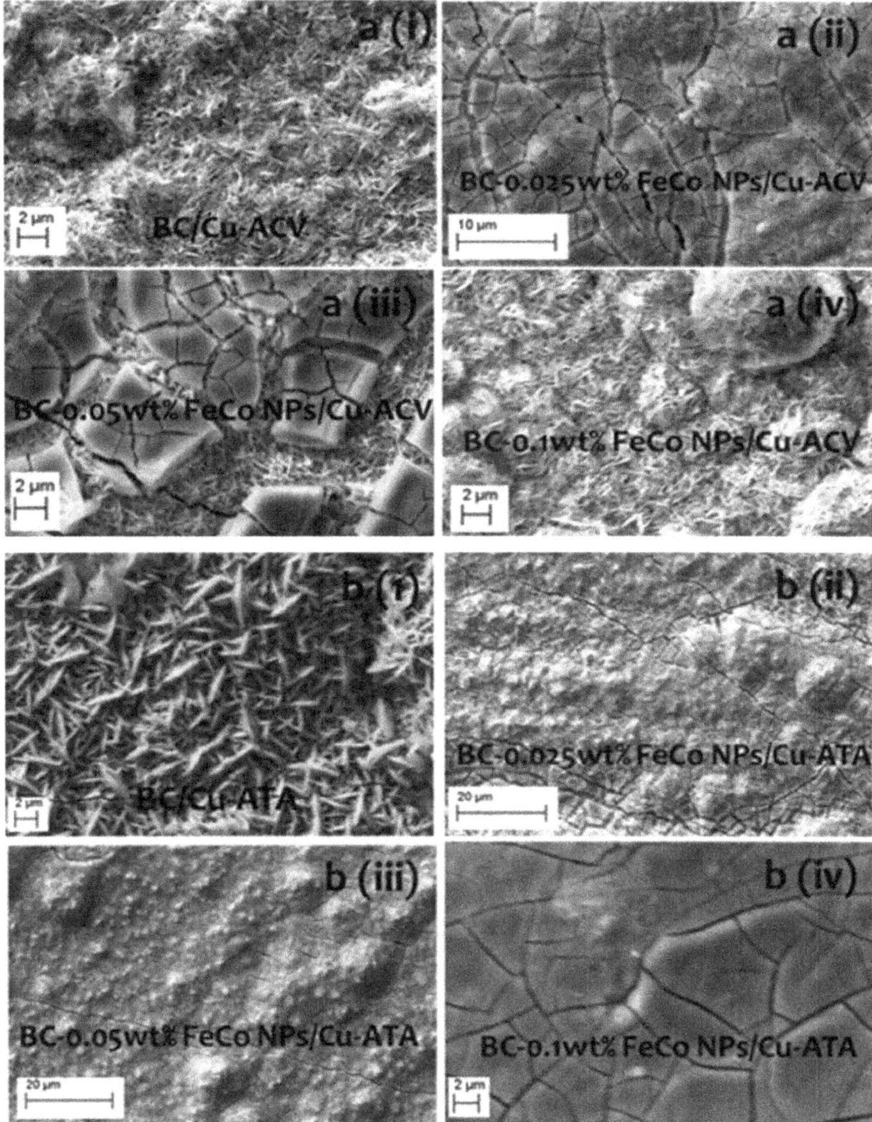

FIGURE 4.16 SEM surface micrographs of (a and b) (i) black chrome and (a and b) (ii, iii, and iv) FeCo(C) NPs modified black chrome selective coatings with varying wt.% of FeCo(C) NPs deposited on copper substrates (left panel) after cyclic voltammetry measurements and (right panel) after corrosion measurements.

in Figure 4.17b. This is due to the partially filled cracks and a mixture of partially smooth and partially denser nano grains regions, as also has been observed in SEM surface microstructure analysis, Figure 4.16 a(ii and iii).

AFM surface morphologies of black chrome and FeCo (C) NPs modified black chrome selective coatings on a copper substrate after corrosion measurements are

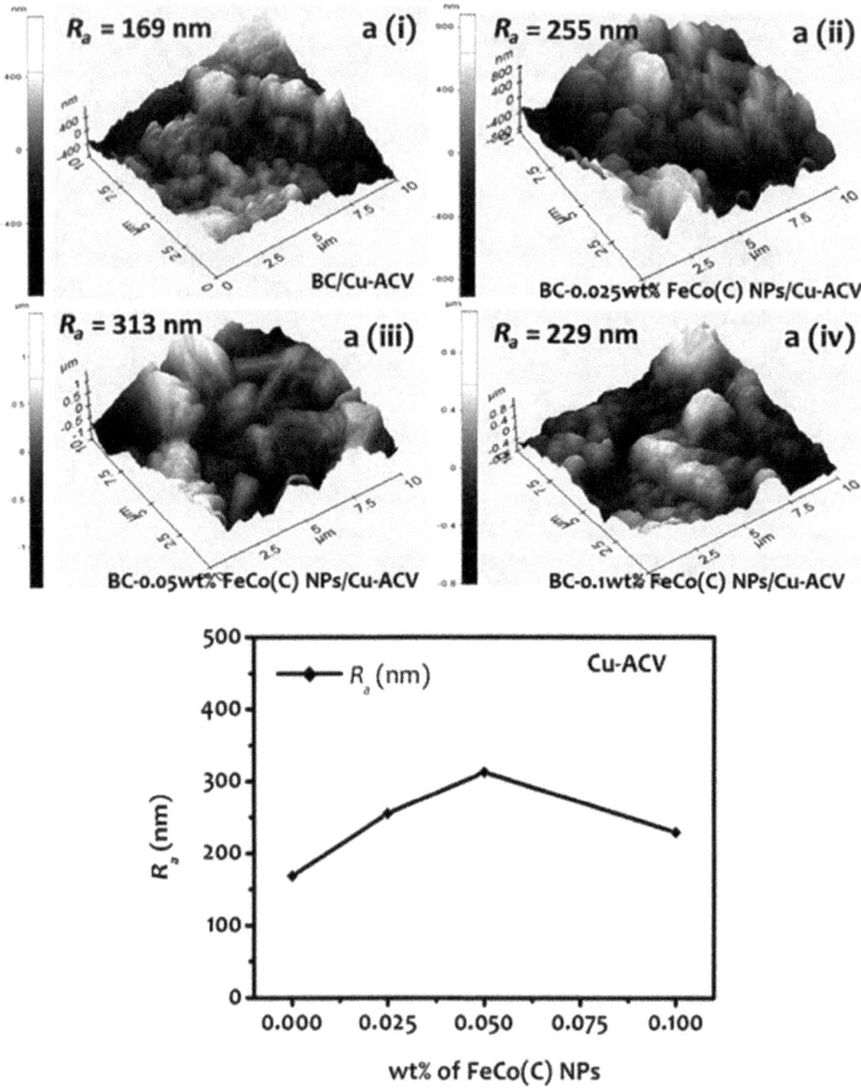

FIGURE 4.17 AFM surface morphology of (a)(i) black chrome and (a)(ii, iii, iv) FeCo(C) NPs modified black chrome selective coatings with varying wt.% of FeCo(C) NPs deposited on copper substrates, after cyclic voltammetry measurements and (b) average surface roughness (R_a) with varying wt.% of FeCo(C) NPs.

shown in Figure 4.18 a(i, ii, iii, and iv). The average surface roughness (R_a) of these coatings has decreased with increasing wt.% of FeCo (C) NPs in modified black chrome selective coatings (Figure 4.18b), which is mainly attributed to the reduction in cracks and other surface irregularities as shown in Figure 4.16 b(ii, iii, and iv), also substantiating the SEM findings, as discussed previously.

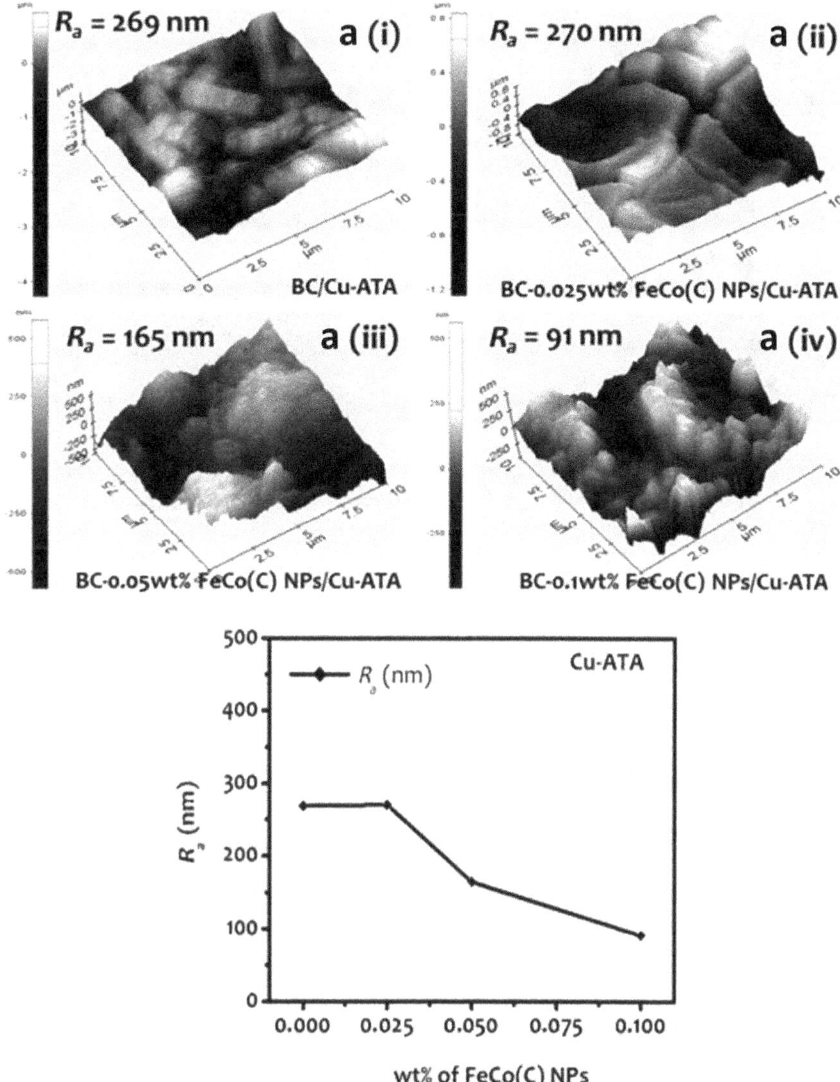

FIGURE 4.18 AFM surface morphology of (a)(i) black chrome and (a)(ii, iii, iv) FeCo(C) NPs modified black chrome selective coatings with varying wt.% of FeCo(C) NPs deposited on copper substrates, after corrosion measurements and (b) average surface roughness (R_a) with varying wt.% of FeCo(C) NPs.

Energy-Dispersive X-ray (EDS) spectra of black chrome and FeCo (C) NPs modified black chrome selective coating on copper substrate are shown in Figure 4.19 a(i, ii, iii, and iv) after cyclic voltammetry (CV) measurements and in Figure 4.20 b(i, ii, iii, and iv) after corrosion measurements. The corresponding elemental compositions in weight% and atomic% are represented as insets in these EDX spectra

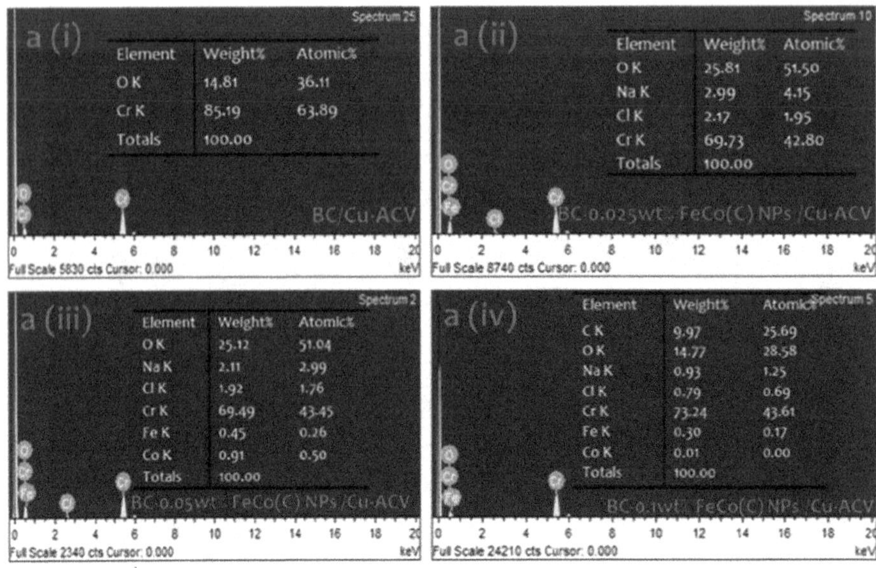

FIGURE 4.19 EDS elemental analysis of (a)(i) black chrome and (a)(ii, iii, iv) FeCo(C) NPs modified black chrome selective coatings with varying wt.% of FeCo(C) NPs deposited on copper substrates, after cyclic voltammetry measurements.

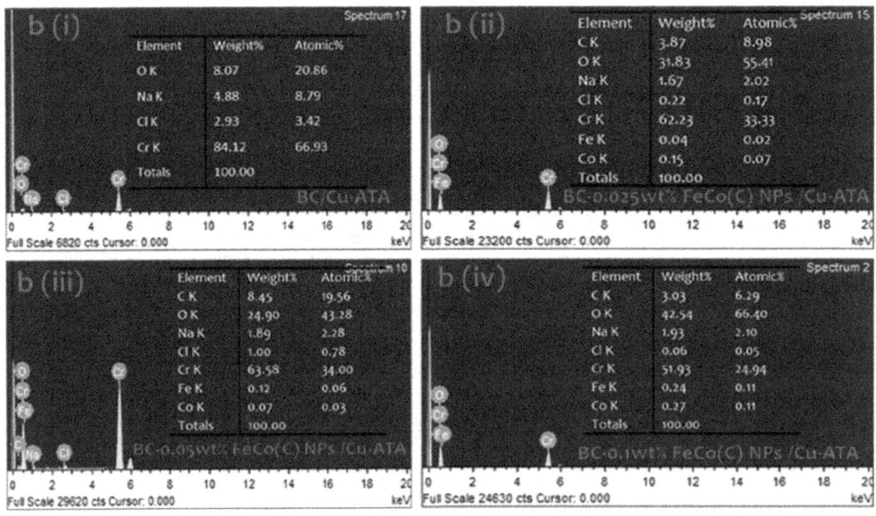

FIGURE 4.20 EDS elemental analysis of (b)(i) black chrome and (b)(ii, iii, iv) FeCo(C) NPs modified black chrome selective coatings with varying wt.% of FeCo(C) NPs deposited on copper substrates, after corrosion measurements.

(Figure 4.19 a(i, ii, iii, and iv) and Figure 4.20 b(i, ii, iii, and iv)). After cyclic voltammetric measurements, oxygen and chromium atomic fractions are 36.11 atm.% and 63.89 atm.% (Figure 4.19 a(i)), respectively. These atomic fractions have changed drastically as compared to 74.58 atm.% and 25.42 atm.% fractions of oxygen and chromium before CV measurements (Figure 4.6(a)). Similar changes have been observed for FeCo(C) NPs modified black chrome coatings after cyclic voltammetry measurements, where, as an example, oxygen atomic fraction has reduced from 64.22 atm.% (before cyclic voltammetry measurements) to 51.50 atm.%, and chromium atomic fraction has increased from 27.63 atm.% (before cyclic voltammetry measurements) to 42.80 atm.% for 0.025 wt.% FeCo (C) NPs modified black chrome coatings. The changes in the atomic fraction of elemental compositions have been summarised in respective tables. In addition to the observed expected elements, some undesired elements have also been observed such as sodium and chlorine in very small atomic fractions in the case of cyclic voltammetry treated coating structures. The presence of these undesired elements may be due to the saline electrolyte solution used for these cyclic voltammetry experiments.

Similar trends for an atomic fraction of different elements have also been observed after corrosion measurements for black chrome and FeCo (C) NPs modified black chrome selective coatings (Figure 4.20 b(i, ii, iii, and iv)). The elemental analysis has been carried out at different regions, and it has been observed that FeCo (C) NPs are distributed homogeneously thus avoiding the oxidation of metallic chromium in FeCo (C) NPs modified black chrome coatings. Thus, the environmental stability of the black chrome has increased relatively after the modification of graphite-encapsulated FeCo nanoparticles (FeCo (C) NPs). Similar observations have been observed in the case of FeCo(C) NPs modified black chrome selective coatings on nickel-seeded copper substrate, after cyclic voltammetry and corrosion measurements and discussed by Usmani *et al.* in detail [Usmani and Harinipriya, 2013].

4.5.2 Optical Properties after Electrochemical Measurements

The measured reflectance spectra in 200–800 nm wavelength range of black chrome and FeCo (C) NPs modified black chrome selective coatings on Cu and Ni/Cu substrates are shown in Figure 4.21 (a and b). These measurements were recorded after cyclic voltammetric measurements on these coating structures and used for calculating absorptance for these structures. The calculated absorptance values are summarised in the respective insets of Figure 4.21 (a and b). The absorptance values are approximately 0.96 for all these structures after CV measurements, especially on Cu substrates, whereas absorptance values are approximately 0.95 on Ni/Cu substrates. The observed changes are really insignificant and are within the error limits of measurements and related calculations.

Figure 4.22 (a and b) summarises reflectance spectra of BC and FeCo (C) NPs modified BC selective coatings in 2.5–25 μm wavelength range, after CV measurements, and used to calculate the emittance values. The measured emittance values

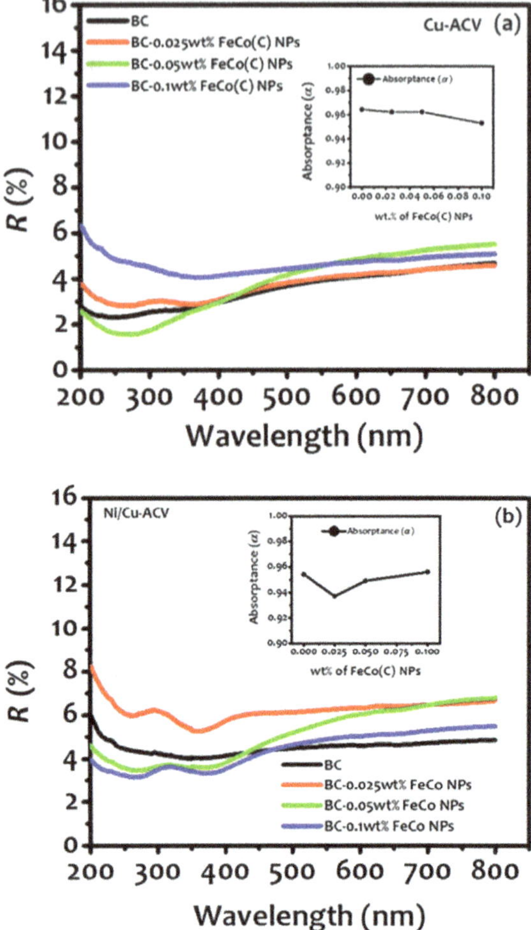

FIGURE 4.21 Reflectance spectra versus wavelength of black chrome and FeCo(C) NPs modified black chrome selective coatings with varying wt.% of FeCo(C) NPs deposited on (a) Cu and (b) nickel-coated Cu substrates, after cyclic voltammetry measurements. Calculated absorptance values versus varying wt.% of FeCo(C) NPs have been represented in inset in the graphs of a and b.

are shown as insets in Figure 4.22 (a and b). These values are relatively on the higher side and are nearly similar to the values for coatings without cyclic voltammetry and corrosion measurements. However, FeCo (C) NPs modified black chrome selective coatings on both copper and nickel-coated copper substrate showed relatively lesser emittance values, after CV measurements. These observations suggest that FeCo (C) NPs modified black chrome selective coatings may be more environmentally stable as compared to the pristine black chrome coatings and thus can be used under saline conditions and at elevated temperatures.

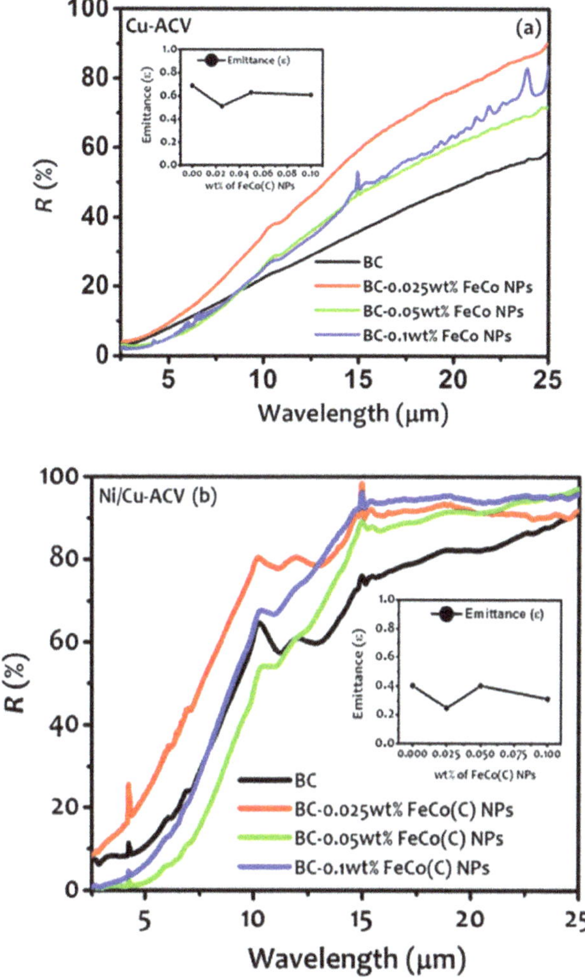

FIGURE 4.22 Reflectance spectra verses wavelength in the range of 2.5–25 μm of black chrome and FeCo(C) NPs modified black chrome with varying wt.% of FeCo(C) NPs selective coatings deposited on (a) copper and (b) nickel-coated copper substrates, after cyclic voltammetry measurements. Calculated emittance values versus varying wt.% of FeCo(C) NPs have been represented in insets in the graphs.

5 High-Temperature Solar Selective Absorber Surfaces

ZrO_x/ZrC-ZrN/Zr Tandem Structure

5.1 INTRODUCTION

Numerous oxide and nitride-based solar selective coatings are investigated for high-temperature applications. A historical development of high-temperature SSCs is discussed in Section 2.4.3. Among them, titanium, zirconium, or hafnium metal carbides, oxides, and nitrides exhibit a higher degree of spectral selectivity, showing promise for spectrally selective coatings, especially for high-temperature applications. These carbides and nitrides absorber-reflector tandem structures such as TiN_x, ZrN_x, and ZrC_xN_y with silver (Ag) as an infrared reflector on different substrates have been investigated as the solar selective coatings for desired solar absorptance and thermal emittance properties [Kennedy, 2002]. These selective absorbers exhibit stability problems at high temperature approximately 350°C (even in a vacuum) due to the agglomeration issues of the silver metal reflector. However, these structures showed good solar thermal performance at room temperature. Poor thermal stability of this metal-reflecting layer at high temperature (~ 350°C) degrades the infrared reflecting properties, causing an enhancement in the emissivity for SSCs. Sputtered ZrC_xN_y selective absorber surfaces on aluminium-coated oxidised stainless steel are thermally stable up to 600°C (likely in a vacuum but not specified in reported work). Lazarov *et al.* have replaced Ag with Zr as an infrared reflector layer in sputtered ZrO_x/ZrC_x/Zr absorber-reflector tandem solar selective structures and observed a high optical selectivity with α/ε (20°C) value of 0. 90/0.05 and thermal stability on stainless steel and quartz substrates up to 600°C and 800°C in vacuum, respectively [Lazarov and Mayer, 1997, 1998]. There are still several challenges associated with such high-temperature solar selective coatings, especially for their long-term thermal stability, cyclability, and environmental stability (corrosion, etc.). Keeping in mind these constraints, continuous efforts are in progress towards the development of such coatings, which may provide an enhanced spectral selectivity and thermal stability to realise the better power-generation efficiencies.

In this study, attempts are made to design and develop a high-temperature absorber-reflector (ZrO_x/ZrC_x/Zr) tandem structure-based spectrally selective coating. This includes: (i) the optimisation of zirconium film as a metal reflector with

DOI: 10.1201/9781003563990-5

minimum emittance, (ii) the optimisation of zirconium carbide-nitride absorber layer with an enhanced solar performance and temperature stability, and (iii) understanding microscopic origin of absorptance, emittance, and other physical properties in such a system. The developed ZrO_x /ZrC_x /Zr structures on copper and stainless substrates were further subjected to numerous characterisations to understand (i) the impact of growth conditions on mechanical properties and (ii) the impact of corrosion and thermal treatment on physical properties and their correlation with the solar thermal performance.

5.2 DESIGN OF ZrO_X/ZrC-ZrN/Zr ABSORBER-REFLECTOR TANDEM STRUCTURES FOR SPECTRALLY SELECTIVE COATINGS

The schematic of $2\theta = 29.89°$ absorber-reflector tandem structure is shown in Figure 5.1(a), for different substrates such as stainless steel (SS), copper (Cu), aluminium (Al), and glass. Here, a metallic Zr layer is used to achieve infrared reflection in the desired wavelength range on these substrates against the conventional infrared reflecting layers such as silver and aluminium, where stability is limited up to approximately 350°C, even in a vacuum. Against these conventional infrared reflectors, Zr metal reflector provides a high infrared reflectivity with an enhanced thermal stability against high temperature because of its refractory nature [Zhang *et al.*, 2003]. In conjunction with the infrared reflector layer, the absorber layer is also critical and needs to be optimised for the maximum absorption of incident solar radiation. ZrC-ZrN absorber layers are investigated and optimised by controlling the free electron density in the d-band of Zr transition metal and nitride fraction in carbide matrix. This can be achieved by manipulating the stoichiometry and nitrogen concentration in ZrC-ZrN absorber layers. The enhanced electron density in ZrC-ZrN spectrally selective absorber layers will introduce plasmonic absorption in the long-wavelength region of the solar spectrum. In conjunction with intrinsic absorption, this additional plasmonic absorption, in absorber layers, will enhance the total absorption of the incident solar radiation over an extended wavelength range [Seraphin, 1979; Kittel, 1995]. We utilised this concept to optimize the optical properties of absorber layers to achieve the maximum spectral selectivity and, thus, the enhanced solar absorption, by varying nitrogen concentration. Antireflection structures are used to minimise the back reflection into the ambient atmosphere and any microstructural damages to the absorber layer directly. In this work, ZrC_x antireflecting layer was fabricated on ZrC-ZrN absorber layer, as explained in Figure 5.1. ZrC_x antireflection layer not only reduces the back reflection but also enhances the thermal and environmental resistivity of fabricated spectrally selective structures. We observed that ZrO_x /ZrC_x /ZrC-ZrN/Zr absorber–reflector tandem layered structures exhibit absorptance values of 0.81–0.88, 0.81–0.90, 0.82–0.90, and 0.86–0.89 and emittance values of approximately 0.04, 0.1, 0.05, and 0.05 on SS, Cu, Al, and glass substrates, respectively. The detailed fabrication process and intensive characterisation of these designed spectrally selective coatings are discussed in the following sections.

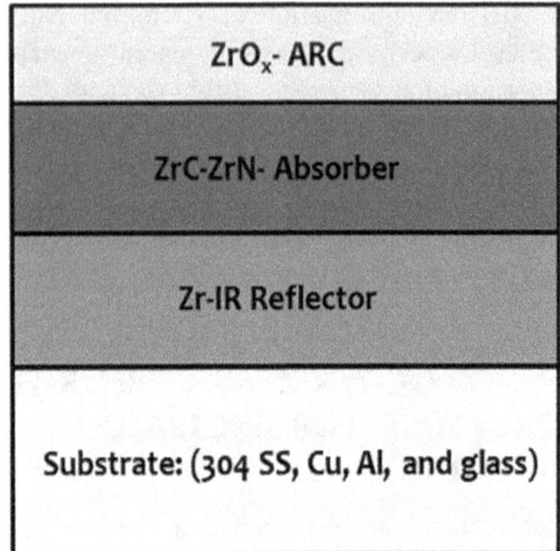

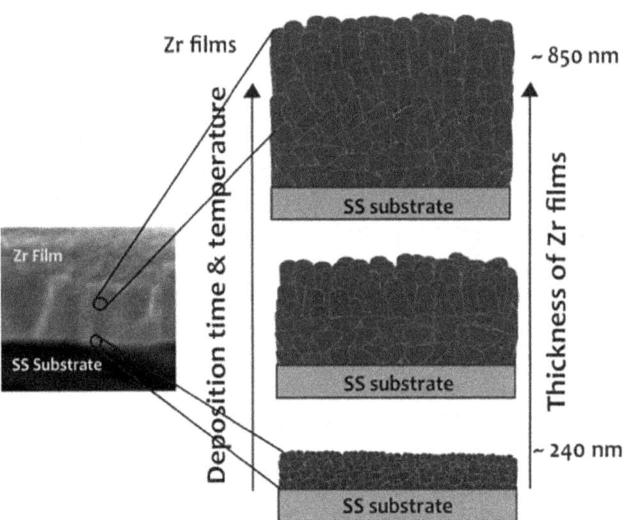

FIGURE 5.1 (a) Schematic representation of ZrO_x/ZrC-ZrN/Zr absorber-reflector tandem structures on stainless steel (SS), copper (Cu), aluminium (Al), and glass substrates. (b) Growth schematic of Zr metallic film on SS substrate with varying deposition time and temperature.

5.3 OPTIMISATION OF SPUTTERED ZIRCONIUM FILM AS AN INFRARED REFLECTOR

The optimisation of zirconium (Zr) layer is important to achieve the desired solar thermal performance. This includes the process parameters and effective thickness, which may provide the minimum thermal emittance. Zirconium thin films were deposited using DC sputtering at a working pressure approximately 2.5×10^{-2} mbar. The argon (Ar) sputtering gas was introduced at a flow rate of 50 SCCM (standard cubic centimetre per minute at STP) into the chamber at the mentioned working pressure for deposition. The process parameters such as time and temperature of deposition have been optimised keeping the power constant. The synthesis processes, including the cleaning of substrates, are discussed in detail by Usmani *et al.* [Usmani *et al.*, 2016b].

5.3.1 Results and Discussion

5.3.1.1 Structural Analysis

X-ray diffraction measurements are shown in Figure 5.2 for Zr/SS structures, deposited for different time intervals and substrate temperatures. X-ray diffraction data

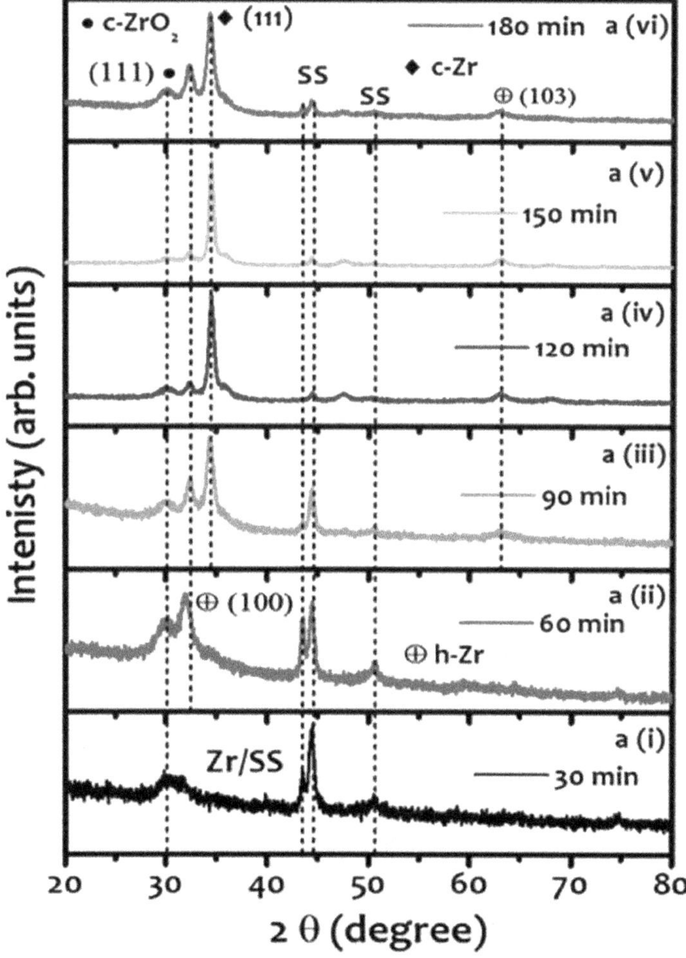

FIGURE 5.2 X-ray diffraction spectra of the Zr films sputtered at different (a) deposition times and (b) deposition temperatures on stainless steel substrates.

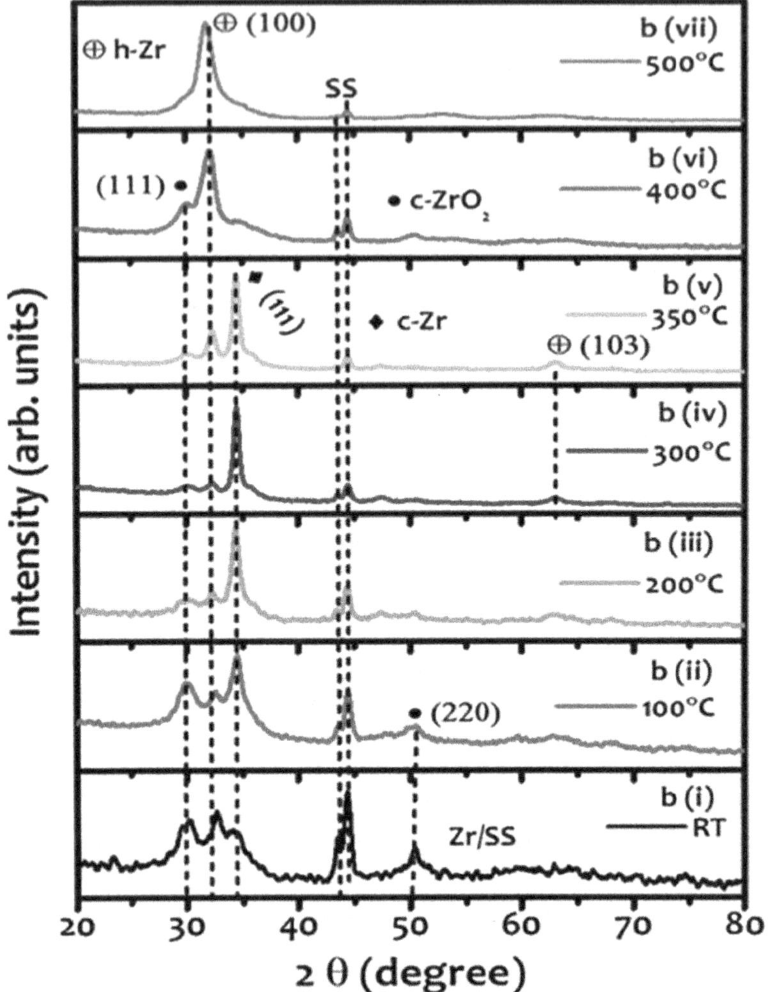

FIGURE 5.2 (Continued)

(Figure 5.2) suggests that the sputtered deposited Zr films are polycrystalline in nature, showing a mixture of cubic and hexagonal zirconium phases. The deposition time has a strong effect on the crystallographic phase, as can be seen from the X-ray diffraction patterns (Figure 5.2a (i, ii, iii, iv, v, and vi)). For 30-min deposition time (low thickness), Zr thin film is nearly amorphous, and only chromium peaks at $2\theta = 43.54°$, $44.44°$, and $50.55°$, from SS substrate (ICDD PDF#: 01–088–2323), can be seen. A very less intense peak at $2\theta = 29.89°$ has been observed, which may correspond to zirconium oxide (111) diffraction plane (ICDD PDF#: 049–1642). The onset of this oxide phase is attributed to the presence of residual oxygen in the deposition chamber. However, for 60-min deposition, the Zr film showed a diffraction peak at 2θ = 31.97° corresponding to (100) plane for hexagonal zirconium crystallographic phase (ICDD PDF#: 03–065–3366), in conjunction with substrate diffraction peaks. As the deposition time of Zr film was increased, Zr film exhibited (111)-preferred cubic crystallographic phase at 2θ = 34.386° (ICDD PDF#: 01–088–2329). The phase development of Zr films on glass substrate also followed the same pattern as on SS substrates with varying deposition times. The substrate temperature may also play an important role in the crystalline quality of sputtered films and has been used as an optimisation parameter for Zr thin films in the present study [Bilgin *et al.*, 2005]. Figure 5.2b shows the X-ray diffraction pattern of Zr films deposited on SS substrate at different temperatures, keeping the deposition time constant for two hours. The films deposited at different temperature are polycrystalline and polyphasic in nature with simultaneous cubic and hexagonal zirconium crystallographic phases. At a low substrate temperature, the film exhibited (111)-preferred orientation of cubic zirconium phase at 2θ = 34.386° (ICDD PDF#: 01–088–2329). However, (100) orientation of hexagonal zirconium film is favoured with increasing temperature from 350°C to 500°C. The oxide impurity starts reducing with increasing both deposition time and temperature, as shown in Figure 5.2, where the observed zirconium oxide diffraction peak has reduced for films deposited at higher temperatures and for longer durations.

The full-width half maxima (FWHM), *B*, grain size (*t*) [Cullity, 1972], dislocation density (σ)[Williamson and Smallman, 1956], and strain (ε) [Velumani *et al.*, 2003] of the films were calculated for the preferential orientations, and results are summarised in Tables 5.1 and 5.2 for the Zr films deposited on SS substrate with

TABLE 5.1

Microstructural properties of Zr films with varying deposition times on SS substrates

Time (min)	*B* (rad)	*t* (nm)	$\delta \times 10^{16}$ (line/m^2)	ε
90	0.711	11.7	0.7305	2.2527
120	0.586	14.2	0.4959	1.856
150	0.516	16.12	0.3848	1.6348
180	0.692	12.01	0.6932	2.1924

TABLE 5.2

Microstructural properties of Zr films with varying deposition temperature on SS substrates

T (°C)	*B* (rad)	*t* (nm)	$\delta \times 10^{16}$ (line/m^2)	ε
100	0.966	8.61	1.3489	3.06
200	0.785	10.6	0.8899	2.4871
300	0.583	14.27	0.4959	1.8471
350	0.638	13.04	0.588	2.021

varying deposition times and substrate temperatures. However, the details of these parameters on glass substrates are described elsewhere [Usmani *et al.*, 2016b]. The crystalline size increases, and the dislocation density decreases with increasing the deposition time and substrate temperature, initially, and thereafter a slight decrease in grain size and an increase of strain and dislocation density are noticed. Moreover, the average grain size increases with increasing deposition time (i.e. with increasing the thickness), but for higher deposition times (i.e. higher thicknesses), the decrease in grain size may be due to the formation of additional smaller grains on the larger grains, as also observed by Velumani *et al.* [Velumani *et al.*, 2003].

Similar trends have been observed for grain size, dislocation density, and strain with increasing the substrate temperature, and measured values are summarised in Table 5.2. The dislocation density and strain are demonstrated by the dislocation network in the films. The decrease in the strain and dislocation density suggests the formation of better-quality films at higher substrate temperatures. Moreover, smaller B (HWFM) values and larger t denote better crystallisation of the films. The minimum emittance values were obtained from Zr films with larger grain size, lower dislocation density, and lower strain values. The schematic of zirconium thin-film growth on stainless steel substrate has been illustrated in Figure 5.1(b), as a function of deposition time and thickness. The zirconium film showed a highly disordered phase at the lower thickness, up to approximately 240 nm, which is formed due to a large mismatch between the lattice parameters of Zr and SS or glass substrate. As thickness increases with time, probably the initial Zr layer acts as a buffer layer for upper Zr thin film structure, thus, resulting in an enhanced crystallinity and reduced defect density, as observed in XRD measurements and summarised in Tables 5.1 and 5.2. Finally, the pure crystallographic phase gets deposited due to the pseudo lattice matching of Zr metallic thin-film structures, which has shown the lowest emittance values, as explained later.

5.3.1.2 Microstructural Analysis

Three-dimensional (3D) AFM surface morphology of Zr films, deposited on SS substrate, at a different substrate temperature (from room temperature (RT) to 500°C) is shown in Figure 5.3 (A, B, C, D, E, F, and G). Films show denser granular morphology, with enhanced granular sizes with increasing the substrate temperature. These results also correlate with the observed X-ray diffraction measurements, as explained in the previous section. Root mean square (RMS) surface roughness (R_q), average surface roughness (R_a), and grain size were calculated from the scanned surface morphology. RMS surface roughness and grain size are shown in Figure 5.4 (a and b) as a function of deposition time and substrate temperature on SS substrate. The RMS surface roughness increases (Figure 5.4 (a)) with increasing the deposition time. However, after a certain deposition time, RMS surface roughness starts decreasing, suggesting that the defects are reduced relatively for thicker films. Moreover, at higher deposition time (180 min), RMS surface roughness starts increasing, suggesting the onset of the surface defects because of local agglomeration or other related synthesis defects. The RMS surface roughness is almost same up to 300ºC with deposition temperature; however, it starts decreasing after 300°C (Figure 5.4 (b)). The grain size of the film increases with temperature up to 200°C initially and remains nearly unaffected from 200°C to 400°C of deposition temperature. This starts increasing again at higher temperatures, which is also consistent with the X-ray diffraction patterns, where FWHM of the observed diffraction has decreased with an increasing temperature. RMS surface roughness and grain size of zirconium films deposited on glass substrate show similar characteristics [Usmani *et al.*, 2016b]. SEM micrographs of Zr films deposited on SS substrate with varying substrate temperatures are shown in Figure 5.3A, B, C, D, E, F, and G. The surface micrographs showed a dense granular microstructure with the substrate imprints. The growth and development of surface morphology are similar to that of AFM measurements. The low surface roughness of the film coatings is important to obtain a low thermal emittance, as emittance relies on the surface properties, especially the smoothness of the surface structure [Duffie and Beckman, 1991].

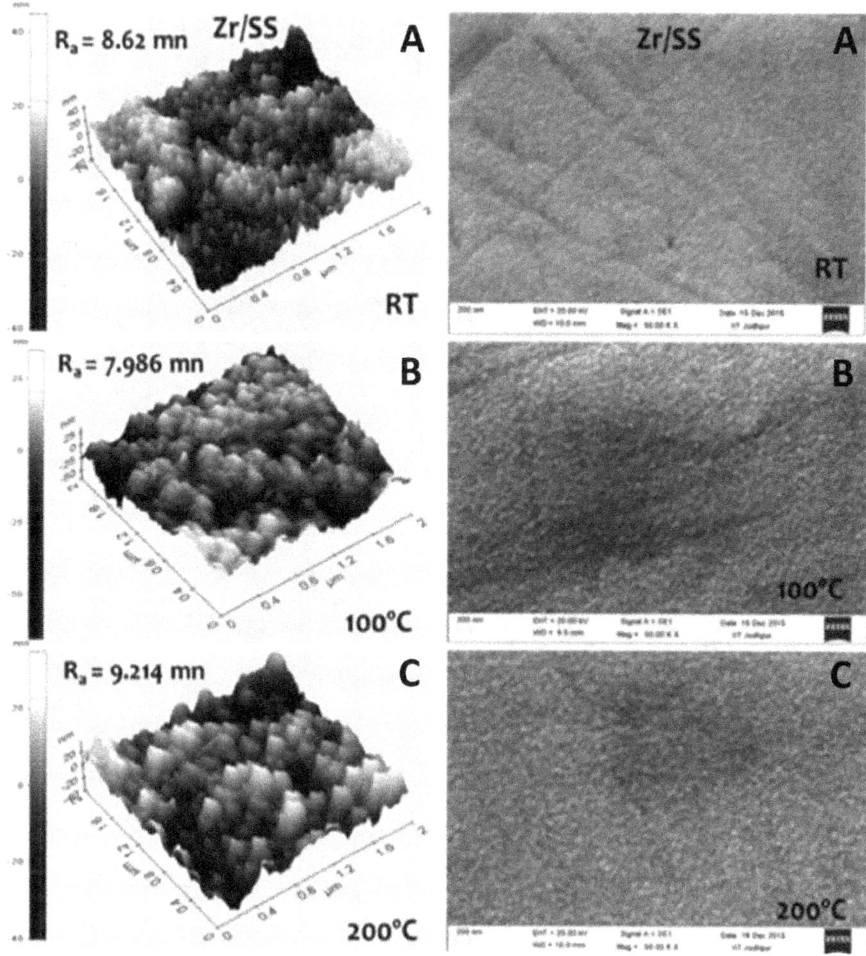

FIGURE 5.3A AND B AFM surface morphology and SEM micrographs images of sputtered deposited Zr thin films on stainless steel substrate having substrate temperatures: (A) RT, (B) 100°C, (C) 200°C, (D) 300°C, (E) 350°C, (F) 400°C, and (G) 500°C.

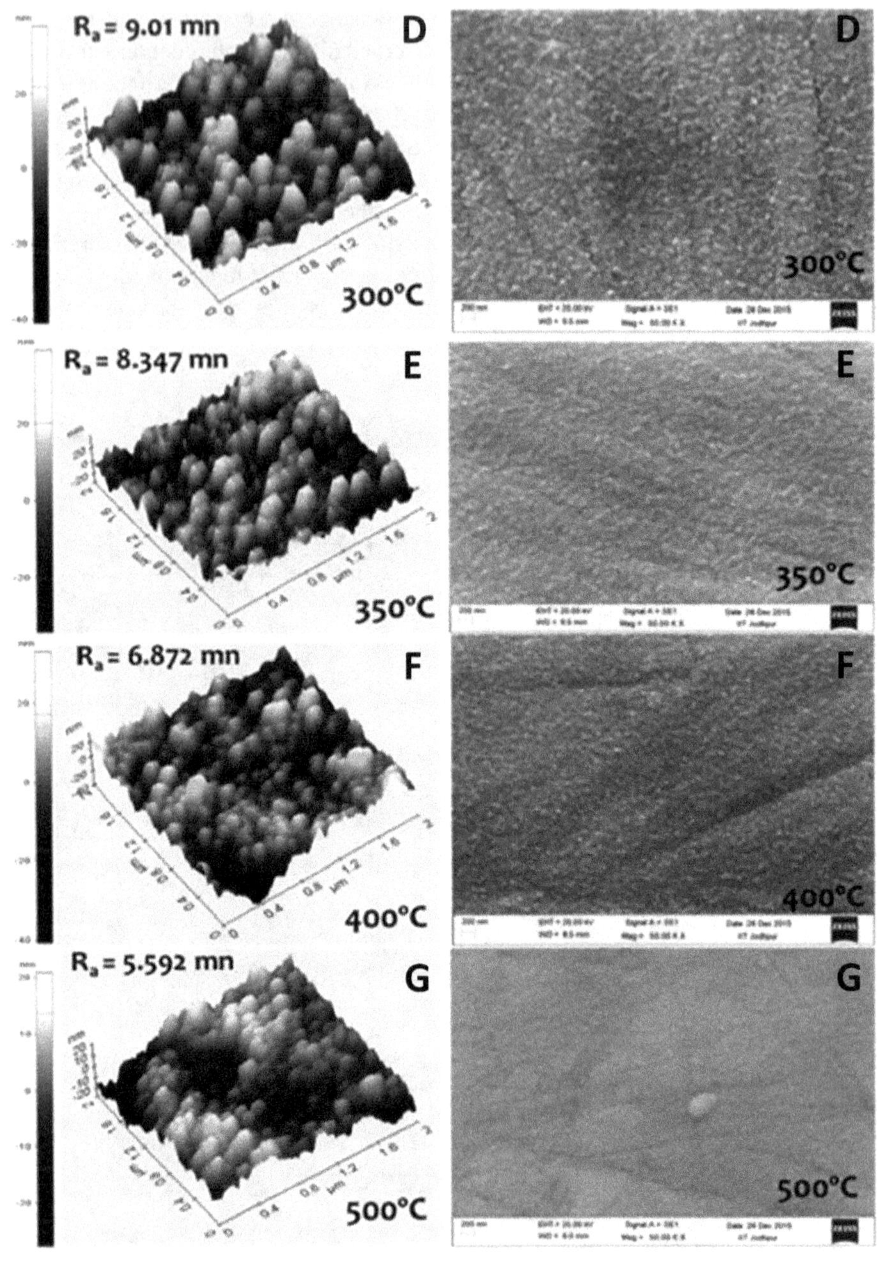

FIGURE 5.3A AND B (Continued)

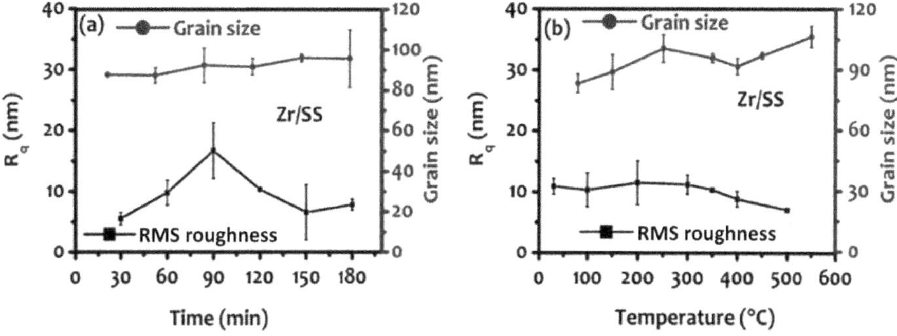

FIGURE 5.4 Root mean square (RMS) surface roughness and grain size of Zr films deposited on SS substrate at different (a) deposition times and (b) substrate temperatures.

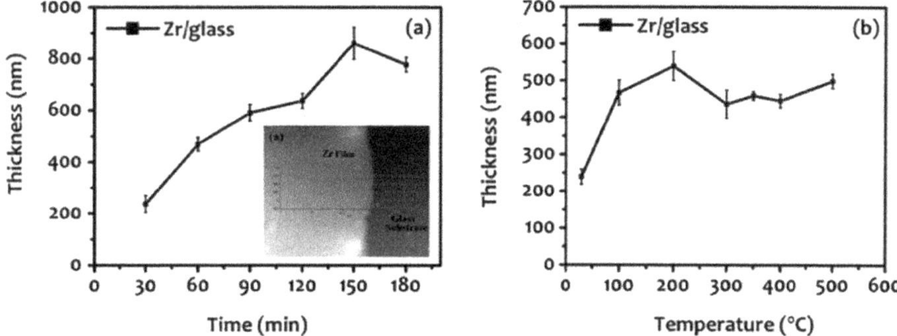

FIGURE 5.5 Film thickness of Zr film on a glass substrate deposited with varying (a) deposition times and (b) substrate temperatures.

The thickness of these films was measured using AFM across the step between Zr film and substrate, as explained in Figure 5.5 (a) inset, and variations in thickness versus deposition time and temperature are plotted in Figures 5.5 (a) and 5.5 (b) respectively. These measurements suggest that film thickness increases continuously; however, it saturates at or above 300°C, as shown in Figure 5.4 (b).

5.3.1.3 Optical Analysis

The reflectance spectra of Zr films are shown in Figure 5.6(a and b) in 2.5–25 μm wavelength range. The calculated emittance values are summarised in the inset of the respective graphs (Figure 5.6a), suggesting that the emittance values decrease with increasing deposition time and thus with the thickness. The minimum emittance value is approximately 0.12, observed for 120 min or larger deposition time. The calculated emittance values of Zr films deposited with varying temperatures on SS substrate are shown in the inset of Figure 5.6(b). The emittance value has decreased from 0.44 to 0.14 with increasing the substrate temperature. The minimum emittance values on SS substrate are observed at 350°C deposition temperature. Thus, the optimal conditions of Zr reflector layer depositions, such as substrate

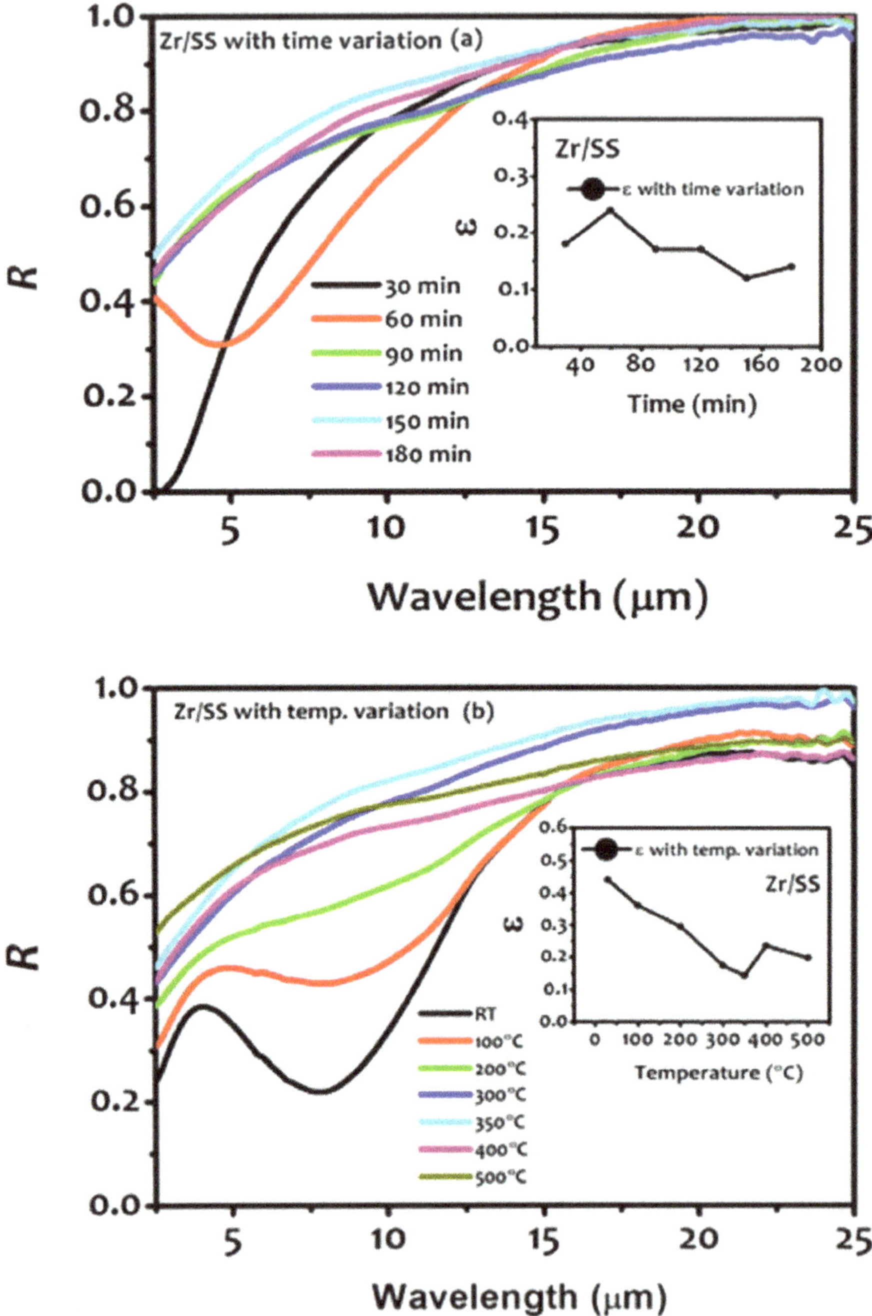

FIGURE 5.6 Reflectance spectra of the Zr film deposited on SS substrate: (a) with changing time from 30 min to 180 min (in inset is shown the plot of calculated emittance versus deposition time) and (b) with changing substrate temperature from room temperature to 500°C (in inset is shown the plot of calculated emittance values versus substrate temperature).

deposition temperature of 350°C and deposition time of two hours, have been used for depositing the Zr metal reflector in tandem ZrOx/ZrC-ZrN/Zr selective structures on different substrates.

5.4 OPTIMISATION OF ZrO_X/ZrC-ZrN/Zr ABSORBER-REFLECTOR TANDEM STRUCTURES

Optimised zirconium metal infrared reflector (IR) was used in zirconium-carbonitride-based absorber-reflector tandem structure (ZrOx/ZrC-ZrN/Zr) as an infrared metal reflector layer. In this section, the details for optimised zirconium carbide-nitride absorber layer on SS and Cu substrates are discussed. The optimised process has also been investigated for Al and glass substrates, and similar results have been observed and discussed in details by Usmani *et al.* [Usmani *et al.*, 2016a]. Spectrally selective tandem absorber-reflector ZrOx/ZrC-ZrN/Zr structures were prepared on different substrates using DC/RF magnetron sputtering system. The samples are labelled as S1 to S6 with increasing nitrogen flow rate used during the synthesis of ZrC-ZrN absorber layers. Details of substrates' cleaning; deposition process; and characterisation techniques such as structural, microstructural, surface, elemental, vibrational, optical properties, and thermal stability are explained in detail by Usmani *et al.* [Usmani and Dixit, 2016a].

5.4.1 Results and Discussion

5.4.1.1 Structural Analysis (X-ray Diffraction and Raman Spectroscopy)

The detailed X-ray diffraction analysis of ZrC-ZrN/ZrOx/ZrC-ZrN/Zr solar selective coatings in sequential deposition order, including substrates, is described in work published by Usmani *et al.* [Usmani and Dixit, 2016a]. We also collected X-ray diffraction spectra of the selective absorber, which is prepared at different nitrogen flow conditions, on both SS and Cu substrates. The representative XRD spectra of solar selective structures on SS substrates are plotted in Figure 5.7a(i, ii, iii, iv, v, and vi) for different nitrogen flow rates used for the fabrication of ZrC-ZrN absorber layers. We noticed similar phases for Zr infrared reflector layers, a ZrC-ZrN absorber layer, and ZrO_x antireflection layers in all these samples with different nitrogen flow rates. X-ray diffraction spectra on Cu substrates are also identical to that on SS substrates. To understand the effect of nitrogen flow rates on ZrN phase evolution, we calculated the relative phase fraction of ZrN in a ZrC matrix. The ZrN phase fraction increased with an initial increase in nitrogen flow and exhibited maxima at 12.5 SCCM, followed by a decrease with any further increase, as explained in Figure 5.7b(i and ii) for layered structures on substrates, stainless steel, and copper.

Room-temperature micro-Raman spectroscopic measurements were carried out for absorber-reflector tandem (ZrO_x /ZrC-ZrN/Zr/Substrate) structure selective coatings and are shown in Figure 5.8. The left panel represents Raman spectra in the range of 200 cm^{-1} to 2000 cm^{-1}, for ZrO_x /ZrC-ZrN/Zr absorber layer, and sequential antireflecting ZrO_x /ZrC-ZrN/Zr layered structures for sample S5 on both the substrates [Figure 5.8(a and b(i)) and Figure 5.8(a and b(ii))]. The observation of characteristic

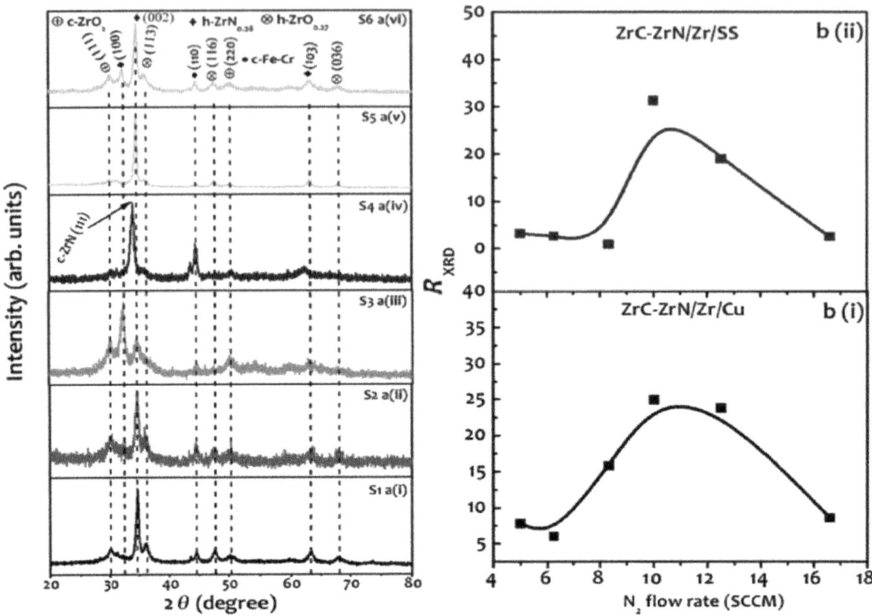

FIGURE 5.7 X-ray diffraction spectra of ZrO_x/ZrC-ZrN/Zr absorber-reflector tandem structures deposited on SS substrate for all samples prepared under different gas flow conditions: a(i) 5 SCCM S1, a(ii) 6.25 SCCM S2, a(iii) 8.34 SCCM S3, a(iv) 10.0 SCCM S4, a(v) 12.5 SCCM S5, and a(vi) 16.6 SCCM S6 and (b) nitride-to-carbide phase fraction with varying nitrogen flow rates in (i) ZrC-ZrN/Zr/Cu and (ii) ZrC-ZrN/Zr/SS absorber layers.

ZrN vibrational modes in ZrC-ZrN absorber layer substantiates the presence of dominating ZrN phase, as observed from XRD measurements also – even for absorber layer with the lowest nitrogen flow rate. The low energy modes approximately 290 cm^{-1} and 360 cm^{-1} represent the characteristic modes for the rock salt ZrN structure, consistent with the reported literature [Han *et al.*, 2005]. The higher wave numbers of approximately 1,350 and 1,580 cm^{-1} vibrational modes show the contribution from C-C and C-N bonding. These numbers are very close to the C-C bonds in intrinsic carbon systems and are found in disordered graphite systems [Kurt *et al.*, 2000; Chowdhury *et al.*, 1999]. The close resemblance of C-C and C-N vibrational frequencies make it difficult to identify the contribution of these vibrational modes in ZrC and ZrN systems independently. Raman spectra for all ZrO_x /ZrC-ZrN/Zr/SS samples from S1 to S6, with different nitrogen flow rates, are plotted in Figure 5.8 (right panel). The results are similar to the observed microscopic changes as explained in previous sections.

The ZrO_x antireflecting layer was deposited on the ZrC-ZrN/Zr layer-coated substrates. The Raman spectrographs of ZrO_x /ZrC-ZrN/Zr multilayer structure for sample S5 are plotted in Figure 5.8 a(ii) and 5.8 b(ii) for SS and Cu substrates, respectively. The observed mode at approximately 610 cm^{-1} is the characteristic mode of the tetragonal ZrO_x system. This is also a characteristic mode for monoclinic ZrO_x system [Siu *et al.*, 1999; Sharma *et al.*, 2014; Keramidas and White, 1974]. The presence

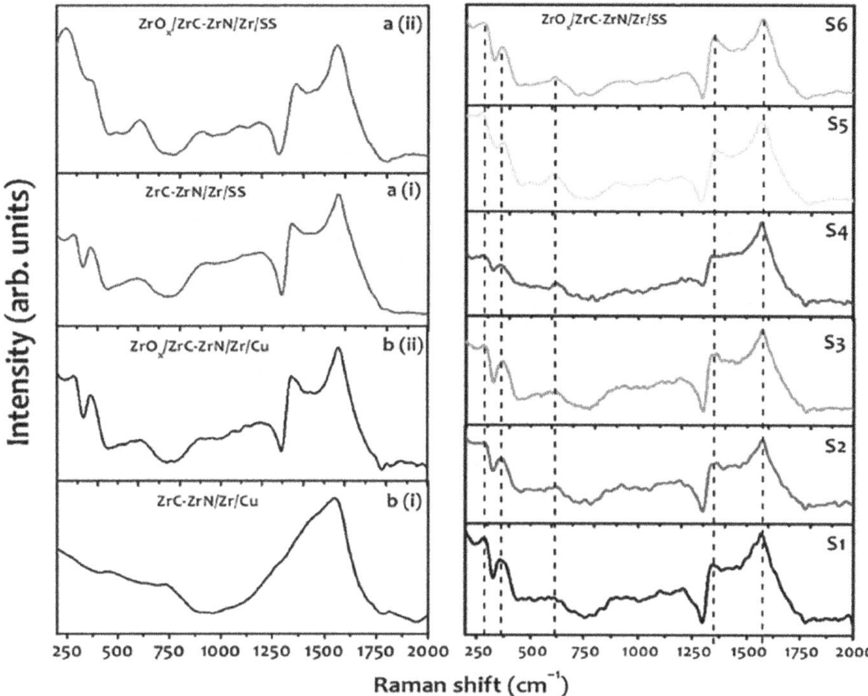

FIGURE 5.8 Left panel: Raman spectrographs for ZrC-ZrN/Zr structures on SS a(i) and on Cu substrate b(i) and for ZrO_x/ZrC-ZrN/Zr structures on SS a(ii) and on Cu substrate b(ii) for S5 sample. Right panel: Raman spectrographs for all ZrO_x/ZrC-ZrN/Zr structures on SS substrates, prepared under different nitrogen flow conditions.

of this characteristic ZrO_x mode suggests the possibility for the admixture of tetragonal and monoclinic phase in room-temperature-grown ZrO_x antireflection layers. The presence of such multiphase is possible for low-temperature-grown ZrO_x thin films. This mode is absent in ZrC-ZrN absorber layer, suggesting the oxide-free ZrC-ZrN absorber layers. The Raman vibrational mode at approximately 1,210 cm^{-1} has been assigned as a second-order Raman mode for 610 cm^{-1} modes for the tetragonal ZrO_x system. The penetration depth (~ 1/2 α, where α is absorption coefficient) of incident 532 nm (~ 2.33 eV) green laser light used in these experiments is approximately 1 μm, much larger than the ZrO_x layer thickness. The large penetration depth of the incident laser light has been observed in terms of a strong interaction with the underneath ZrC-ZrN absorber layer, where all Raman modes, observed in case of pristine absorber layers, are also present in conjunction with ZrO_x vibrational modes.

5.4.1.2 Microstructure, Surface Roughness and Elemental Analysis

The microstructure, roughness, and elemental compositions are analysed using SEM and AFM measurements on these solar selective coatings. The scanning electron micrographs of ZrO_x top surfaces for sample S5 on both SS and Cu substrates are shown in Figure 5.9 (A and B) (b part). We observed that the developed surfaces are smooth with

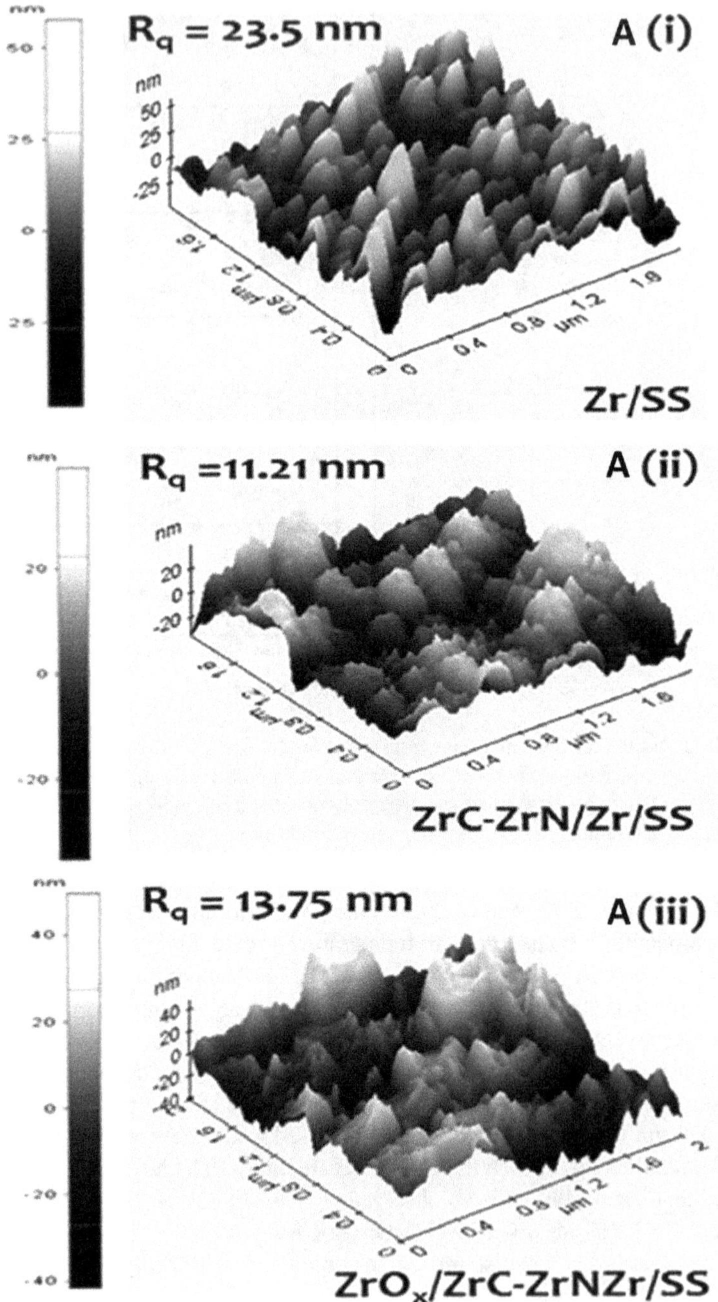

FIGURE 5.9 a: three-dimensional AFM images of Zr layers on a(i) SS substrate, ZrC-ZrN layer deposited on a(ii) Zr-coated SS substrate, and ZrO_x/ZrC-ZrN/Zr coating deposited on a(iii) SS substrate for S5 sample. b: SEM micrographs of the top surface of ZrO_x/ZrC-ZrN/Zr absorber-reflector tandem structure on (a) SS and (b) Cu substrate.

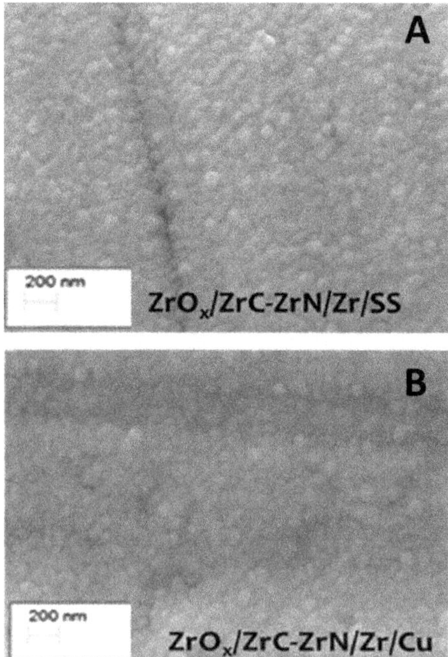

FIGURE 5.9 (Continued)

some substrate surface imprints, developed with the deposition of thin films. These surface imprints are not desired and may degrade the solar thermal response, especially the emissivity of fabricated tandem absorber-reflector structures. In addition, in such surface imprints, developed surfaces show micro-structured columnar grains, as observed in AFM micrographs, and explained in the three-dimensional AFM micrographs (Figure 5.9A(iii) (left panel)] for sample S5 on SS substrate. In addition to the top ZrO_x surfaces, we also collected three-dimensional images of individual, Zr infrared reflector and ZrC-ZrN absorber layers after sequential fabrication, to understand the evolution of microstructures in these solar selective coatings and detailed atomic force micrographs are shown in Figure 5.9 (a). The surface roughness of Zr metal layer is maximum approximately 23.5 nm with hillock-like morphological patterns. This starts reducing with sequential layer deposition, and observed surface root mean square (RMS) roughnesses are approximately 11.21 nm and 13.75 nm with island-type morphological structures for ZrC-ZrN absorber and ZrO_x antireflecting layers, respectively. Similar observations have also been observed for ZrO_x /ZrC-ZrN/Zr/Cu spectrally selective coatings structures and details are present in the published work of Usmani *et al.* [Usmani and Dixit, 2016a].

The elemental analysis has been carried out using EDX measurements for Zr, C, and N atomic fractions on these fabricated ZrC-ZrN absorber layers as a function of N_2 flow rate. The results are summarised in Table 5.3 for SS and Cu substrates. We observed the respective variation in Zr, C, and N atomic fractions by varying N_2

TABLE 5.3
Measured elemental atomic fractions in ZrC-ZrC absorber layers for different samples with respective nitrogen flow conditions used for the synthesis of this structure

		ZrC-ZrN/Zr/SS			ZrC-ZrN/Zr/Cu		
Sr. No.	N_2 (SCCM)	Zr (atm. %)	C (atm. %)	N (atm. %)	Zr (atm. %)	C (atm. %)	N (atm. %)
S1	5.0	26.93	36.42	36.65	21.97	37.12	40.91
S2	6.25	25.56	32.53	41.91	24.97	33.02	42.01
S3	8.34	28.67	27.33	44.0	28.76	25.54	45.70
S4	10.0	35.18	24.09	40.37	30.70	30.13	39.17
S5	12.5	35.98	27.67	36.36	34.14	28.25	37.61
S6	16.66	32.25	27.01	40.65	30.02	29.89	40.09

flow during the synthesis of these tandem absorber-reflector structures. We found that Zr atomic fraction is slowly increasing and showed the maximum Zr atomic fraction for sample S5 at 12.5 SCCM N_2 flow rate. The additional increase in N_2 flow rate resulted in the reduction of Zr atomic fraction. However, we noticed that N atomic concentration is approximately 36%, much larger with respect to that of carbon approximately 27% in the case of S5 sample. Moreover, nitrogen atomic fraction initially increases with the increasing flow rate and starts decreasing at a higher flow rate, which correlates with X-ray diffraction analysis of ZrN fraction in ZrC matrix (Figure 5.7b).

5.4.1.3 Optical Properties

The reflectance versus wavelength measurements are summarised in Figure 5.10 (A and B) for ZrO_x/ZrC-ZrN/Zr structures in approximately 0.3 μm to 25 μm wavelength range on SS and Cu substrates respectively. These are used to calculate the absorptance and emissivity, using Equations (2.5) and (2.7), respectively. The calculated solar absorptance varies by approximately 0.81 to 0.88, and emissivity varies by approximately 0.04 to 0.2 with the nitrogen flow, and results are plotted in Figure 5.10(C) and 5.10(D) for SS and Cu substrates, respectively. In the case of SS substrates, we found that the absorptance is nearly constant having the values of approximately 0.88 up to 8.34 SCCM nitrogen flow rates during deposition, followed by a reduction in absorptance from 0.88 to 0.81 with increasing nitrogen flow rates up to 12.5 SCCM. A similar behaviour was also observed for emissivity, and results are summarised in Figure 5.10(C). The variation in emissivity of these samples is a bit random in the case of structures fabricated on Cu substrates with respect to that of on SS substrates. This may be due to the surface defects and defects in layered structures, which strongly affect the emissivity of the fabricated structure. This is in accordance with the observed high defect density in the case of Cu substrate with

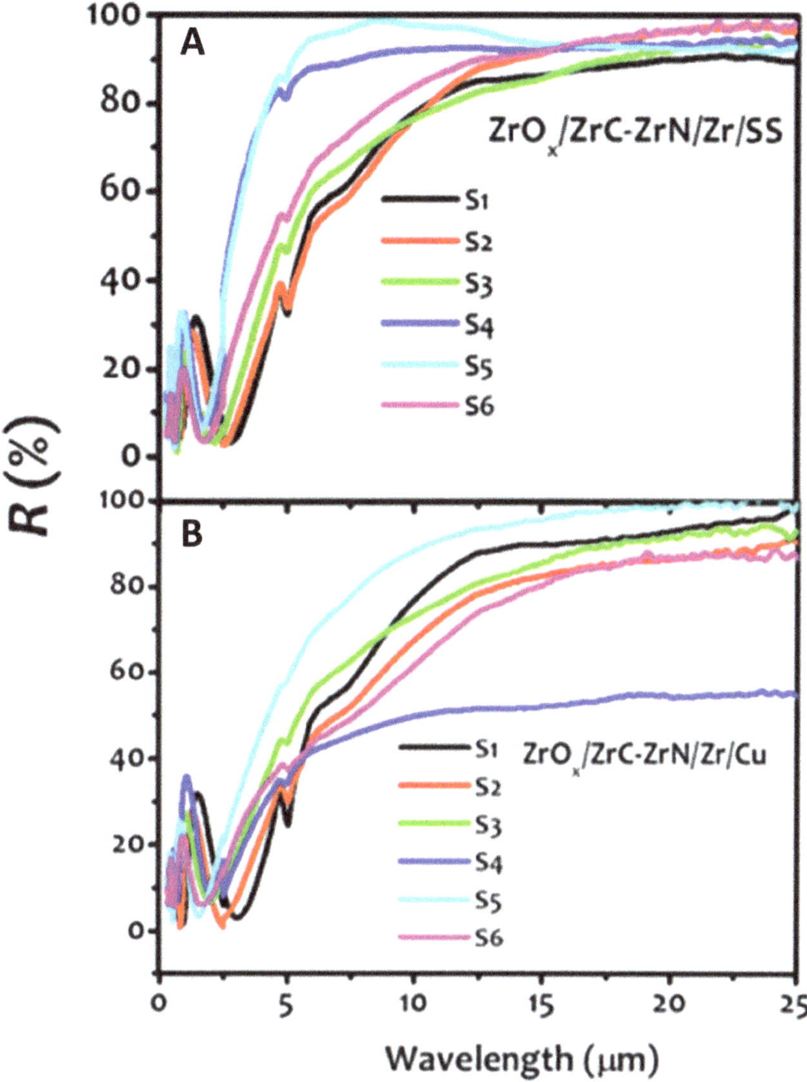

FIGURE 5.10 % reflectance of ZrO_x/ZrC-ZrN/Zr absorber-reflector tandem structures versus wavelength for different samples on SS (A) and Cu (B) substrates; calculated absorptance (α) and emittance (ε) values of these ZrO_x/ZrC-ZrN/Zr absorber-reflector tandem structures versus nitrogen flow rates for SS (C) and Cu (D) substrates.

respect to that of SS substrates, as also observed from Raman vibrational investigations and other microstructural studies, explained previously.

The Kubelka–Munk model was used to calculate the absorptance using $\alpha = \frac{(1-R)^2}{2R}$, where α is absorption coefficient and R is reflectance at a wavelength λ

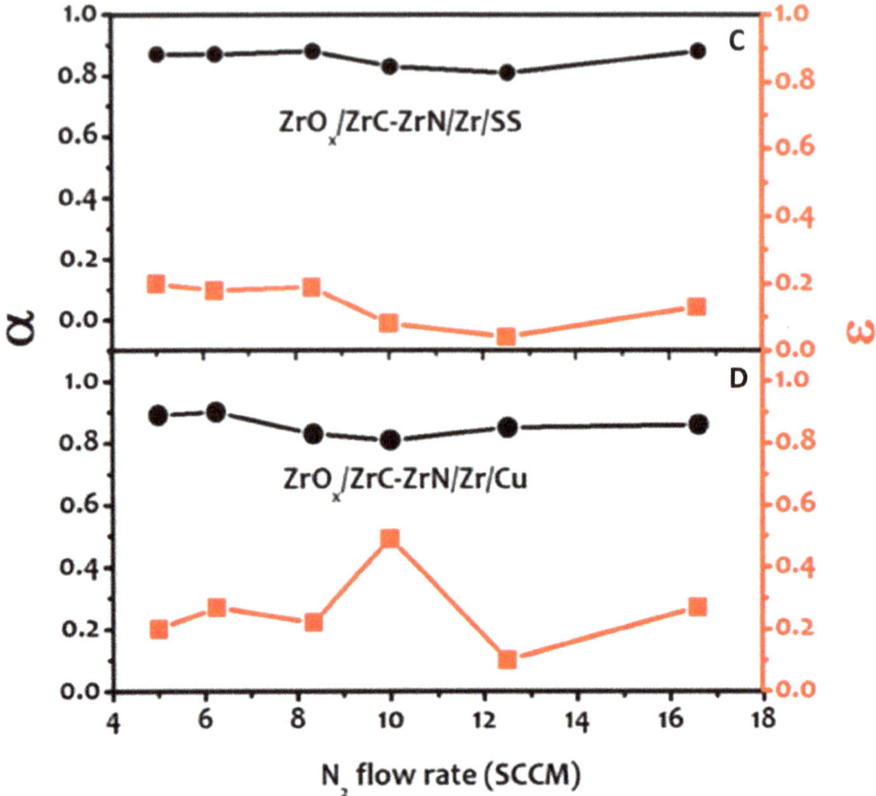

FIGURE 5.10 (Continued)

[Kubelka and Munk, 1931]. The measured normalised absorptances for S5 samples on SS and Cu substrates are plotted in Figure 5.11(A) and 5.11(B) over the entire solar spectrum 0.3–2.5 μm range. Interestingly, we observe three main absorption regions: (i) low wavelength approximately 0.3–0.4 μm region, due to the density of free electrons in the d-band of transition metal [Seraphin, 1979]; (ii) intermediate wavelength approximately 0.4–0.8 μm, due to electronic band gap absorption of the absorbing layer; and (iii) long wavelength approximately 0.8–2.4 μm, due to the excess conduction band electrons in absorbing layer, causing plasmonic absorption. This plasmonic absorption window can be tailored using the nitrogen variation during the synthesis of these solar absorber structures and can be monitored by measuring the electron concentration in these layers.

The number of electrons contributing to the plasmonic absorptions in these samples was calculated from $\omega_p^2 = \dfrac{ne^2}{\varepsilon_0 m}$ where ω_p is plasmon frequency, ε_0 is free space permittivity, and n is the electron density [Kittel, 1995]. The measured plasmon frequencies are plotted in Figure 5.11(C) and (D) for SS and Cu substrates, as a function of nitrogen flow rate, used for fabrication of these layers. We observed an increase

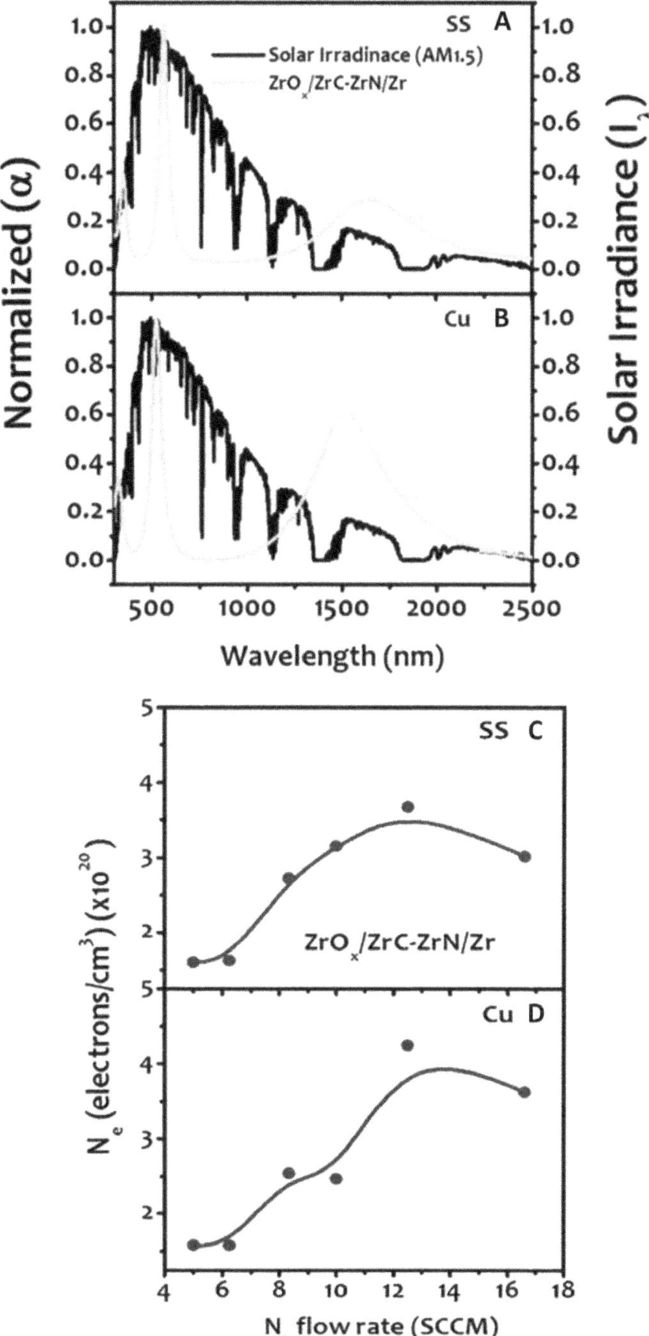

FIGURE 5.11 Normalised absorptance (α) versus wavelength of ZrO_x/ZrC-ZrN/Zr structure on SS (A) and Cu (B) substrates for S5 sample, in conjunction with AM1.5 solar irradiance as a function of wavelength and estimated electron concentration in ZrC-ZrN absorbing layer versus nitrogen flow rates for absorber structures on SS (C) and Cu (D) substrates.

in electron density initially with nitrogen flow rate and maximum electron concentration approximately 4.0×10^{20} cm^{-3} at approximately 12.5 SCCM nitrogen flow rate during synthesis. Further increase in nitrogen flow rate resulted in the reduction of electron concentration in these solar absorber layers. Thus, an optimal nitrogen flow rate is important to achieve the maximum electron concentration, thus plasmon absorption, for enhanced long-wavelength absorption in ZrC-ZrC absorbing layers. We observed that S5 samples on SS and Cu substrates, with the maximum electron concentration, exhibit maximum absorption and lowest emissivity, thus the maximum solar selectivity. These results suggest a novel approach of introducing plasmonic absorption in the absorber layer to tailor the desired absorption in the infrared (IR) wavelength range.

In conjunction with these absorption mechanisms, interference-induced absorption also plays an important role in optimising the solar absorption in multilayered selective structures. The interference-induced absorption relies on refractive index and thickness of the individual layers in the fabricated multilayer structure and substrate properties, whereas intrinsic absorption depends on the extinction coefficient of these layers [Selvakumar *et al.*, 2010; Xinkang *et al.*, 2008]. The normal reflectance measurements are used to calculate the wavelength dependence of refractive index and extinction coefficient using following formulas [Griffiths, 1999].

$$n = \frac{1 + R^{1/2}}{1 - R^{1/2}} \qquad \text{Eq. (5.1)}$$

$$k = \frac{\alpha\lambda}{4\pi} \qquad \text{Eq. (5.2)}$$

Where n is the refractive index, k is extinction coefficient, and R is the reflectance at wavelength λ.

The calculated dispersion curves, that is effective refractive index and effective extinction coefficient as a function of wavelengths, are plotted for a combination of the sequential layers of the S5 sample in Figure 5.12(A) and 5.12(B) (a) on SS substrate. The respective dispersion curves for the Zr metal layer are shown as an inset in Figure 5.12 (A and B) (a) in 200–800 nm wavelength range, and an increase in 'n' and 'k' with the wavelength substantiate the metallic nature and rules out the oxidation of Zr layer. This is important to harness the infrared reflecting properties of the Zr metal layer in multilayer absorber-reflector tandem structures. The measured 'n' and 'k' for ZrC-ZrN/Zr absorber-reflector and ZrC-ZrN/Zr antireflecting coating in conjunction with absorber-reflector structures exhibit sinusoidal variation with wavelength, suggesting the mixed metal–dielectric characteristics of ZrC-ZrN absorber layer. Du *et al.* have also observed similar observation [Du *et al.*, 2011]. The observed metal–dielectric characteristics are consistent for ZrC-ZrN absorber layer, where ZrC is exhibiting metallic character in ZrN dielectric matrix. In contrast to the sinusoidal nature of effective refractive index, extinction coefficient exhibits a different behaviour and two distinct resonance peaks, responsible for absorption in the layered structure are observed. The low wavelength peak approximately 0.4–0.6 μm is due to the optical band absorption from ZrN semiconducting dielectric films and is responsible for absorption in the high-energy range of the solar spectrum, whereas the long-wavelength broad peak approximately 1.2–1.5 μm is due to the collective oscillations of free electrons (plasmon absorption) in ZrC-ZrN matrix, mainly coming from ZrC metallic system. These both peaks exhibit redshifts after introducing a ZrO_x antireflecting layer on the ZrC-ZrN absorber layer. The shift in optical band absorption peak is because of the combined effect of ZrN and ZrO_x optical band absorptions, where low-energy optical absorption is dominating. In contrast to optical band absorption, a red shift in Plasmon absorption is mainly due to the compensation of free electrons in an absorbing layer with the ZrO_x antireflecting layer. In addition, the relatively large extinction coefficient over the entire solar spectrum for the investigated structure confirms the enhanced solar absorption and, thus, enhanced solar performance. The calculated dispersion curves, that is effective refractive index and effective extinction coefficient as a function of wavelengths of

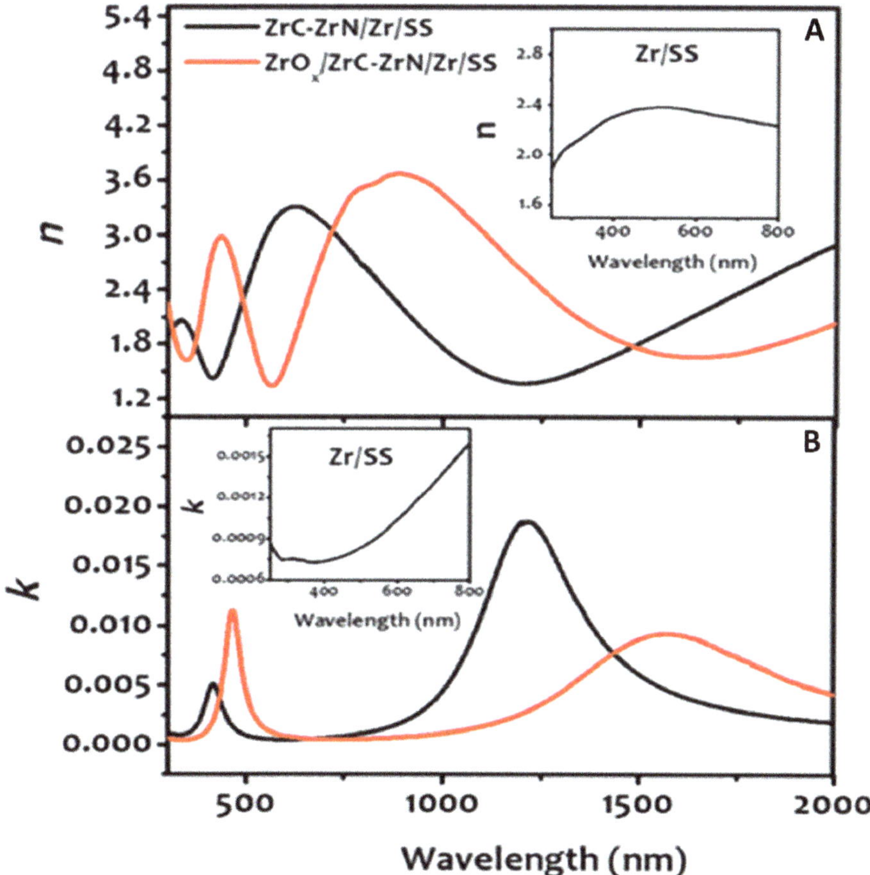

FIGURE 5.12 a: Effective refractive index and effective extinction coefficient versus wavelength of sequential layers for ZrO_x/ZrC-ZrN/Zr absorber-reflector tandem structures fabricated on SS **(A)** and **(B)** substrate. b: Percentage weight change and first-order derivative of weight change as a function of temperature for ZrO_x/ZrC-ZrN/Zr absorber-reflector tandem structures on SS **(A)** and **(B)** substrate in air and nitrogen environment.

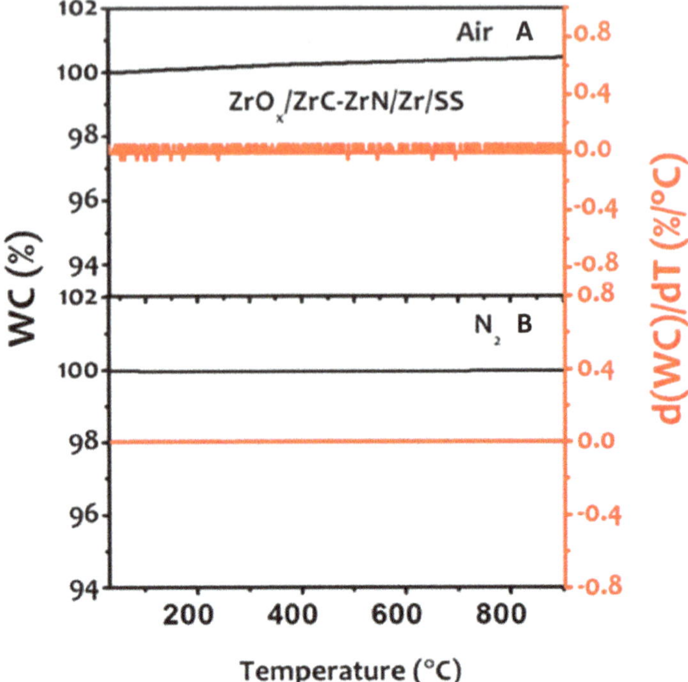

FIGURE 5.12 (Continued)

S5 samples on Cu substrate, show similar behaviour and are discussed in detail by smani *et al.* [Usmani and Dixit, 2016a].

5.5 MECHANICAL PROPERTIES OF ZrO_X/ZrC-ZrN/Zr ABSORBER-REFLECTOR TANDEM STRUCTURES

In this section, we will present the detailed studies of the effect of mechanical properties, such as Young's modulus and hardness, of the synthesised ZrO_x /ZrC-ZrN/Zr absorber-reflector tandem structures on SS and Al substrates. The details about the synthesis of ZrO_x /ZrC-ZrN/Zr spectrally selective reflector-absorber tandem structures are explained earlier and the details are also available in Usmani *et al.* [Usmani and Dixit, 2016a]. Mechanical properties of ZrO_x /ZrC-ZrN/Zr structures are investigated using the nanoindentation technique. The nanoindenter, used for mechanical measurements, is an integrated accessory with atomic force microscopy system (details are given in Section 3.3.9). The Berkovich indenter, a sapphire cantilever with a diamond tip, has been used for indentation on the top surface of ZrO_x /ZrC-ZrN/Zr structures. This indenter was initially forced into ZrO_x /ZrC-ZrN/Zr surfaces at constant load with the forward (down) speed approximately 0.3 μm/s and was kept for ten s at that depth, followed by unloading at a backward (up) speed approximately 0.3 μm/s. Such ten indentations were made on each sample at different physical locations to measure the average hardness and Young's modulus using the generated load–displacement curves. AFM was operated in a contact mode for all such measurements over 5 × 5 μm² scan areas. The details of calculating the hardness and Young's modulus values are also explained in Section 3.3.9. In brief, the hardness is obtained as a function of indent depth during loading by dividing the applied load by the projected contact area of the indenter tip. Poisson's ratio for samples used is approximately 0.3 [Shackelford and Alexander, 2001].

5.5.1 Results and Discussion

Nanoindentation method has been used to measure the hardness and Young's modulus of ZrO_x /ZrC-ZrN/Zr structures, as the thickness of the entire structure is approximately 600 nm, much larger than the nanoindentation depth of approximately 50 nm. This thickness is sufficient to exclude the substrate effect on these measured mechanical properties, and thus measured values may represent the values from ZrO_x /ZrC-ZrN/Zr structures only. The measured values of hardness and Young's modulus are summarised in Figure 5.13 for both SS (a) and Al (b) substrates, as a function of nitrogen flow used for the synthesis of ZrC-ZrN absorber layer. The hardness values show a strong variation for ZrO_x /ZrC-ZrN/Zr structures, with lower nitrogen flow rates used for ZrC-ZrN layer and become nearly constant at ten or higher SCCM nitrogen flow rates, for both the substrates. The hardness values are prone to the film properties such as film density, microstructure, and surface roughness [Chen *et al.*, 2004]. Thus, samples with ten SCCM or higher nitrogen flow may be showing less surface roughness with optimal density of ZrC-ZrN composite absorber, causing nearly constant hardness values in contrast to lower nitrogen flow rates, where the effective density of ZrC-ZrN may be lower and also surface properties may be poor, leading to large variation in hardness values. The maximum selectivity of ZrO_x /ZrC-ZrN/Zr tandem structure selective absorber was found around 12.5 SCCM nitrogen flow rate, used for ZrC-ZrN absorber layer as discussed in the earlier section, suggesting that moderate values of hardness may relatively be better

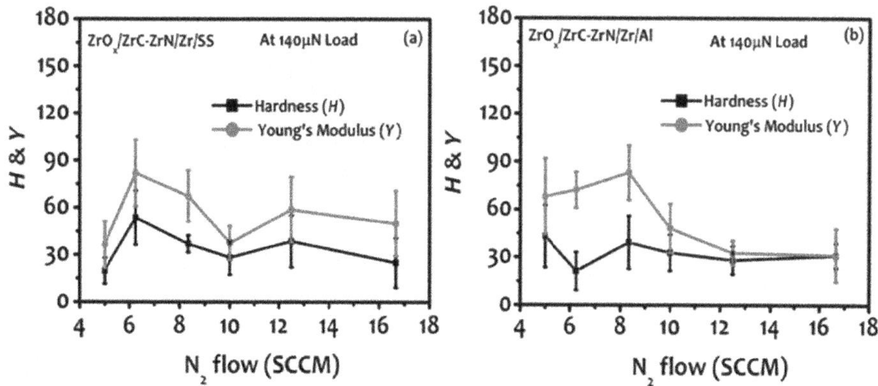

FIGURE 5.13 (a) Hardness (H) and Young's modulus (Y) of ZrO_x/ZrC-ZrN/Zr tandem absorber film on SS substrate and (b) on Al substrate.

for solar thermal performance. Young's modulus also exhibits the similar trend like hardness values, which nearly saturates toward higher nitrogen flow rates, in conjunction with initial large fluctuations.

Hardness and Young's modulus values of ZrO_2, ZrN, and ZrC single-layer thin films are 13.5–19 GPa, 13.4–23.5 GPa, 27.6 GPa and 197–210 GPa, 166.4–196.3 GPa, and 228 GPa, respectively [Chen *et al.*, 2004; Craciun *et al.*, 2010; Zhao *et al.*, 2009]. Interestingly, ZrO_x /ZrC-ZrN/Zr tandem structures exhibit higher values of hardness as compared to that of ZrO_2, ZrN, and ZrC single-layer thin film hardness values. However, Young's modulus values show contrary behaviour with relatively smaller values as compared to ZrO_2, ZrN, and ZrC single-layer thin film values. This is possible as hardness and Young's modulus in the present work represent the effective values of multilayer tandem structures in contrast to the single layer independently.

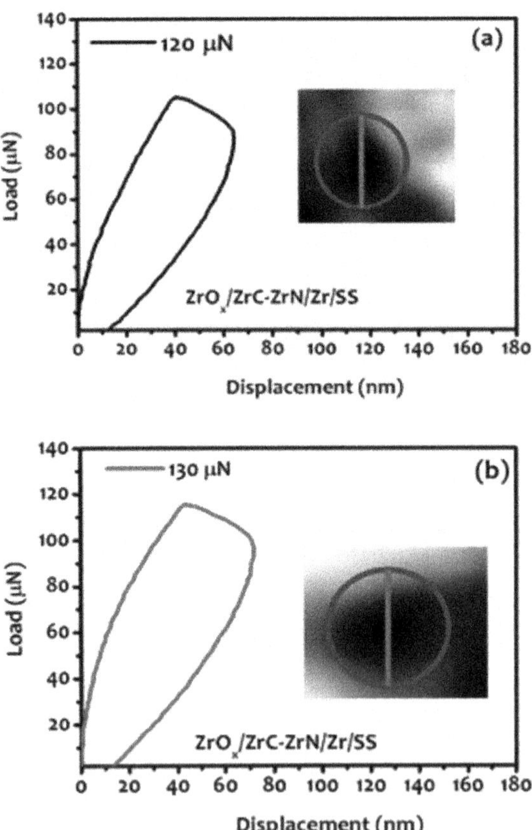

FIGURE 5.14 The load versus displacement curve at difference loads (a to e) with inset showing the respective AFM indentation profiles and hardness and Young's modulus values versus load (f) for ZrO_x/ZrC-ZrN/Zr tandem structure with ZrC-ZrN absorber layer synthesised at 12 SCCM nitrogen flow.

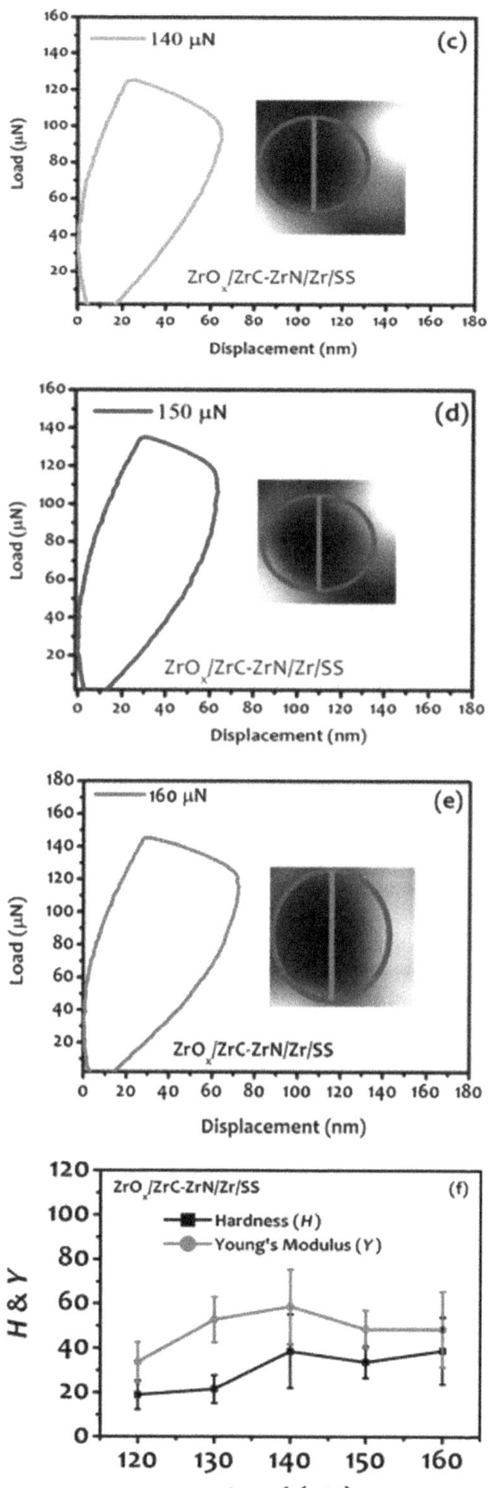

FIGURE 5.14 (Continued)

To further understand the details of mechanical properties, as a function of the load as an external variable, we carried out load versus displacement curve analysis at different initial load conditions on the sample with the optimal solar thermal properties of α approximately 0.81 and ε approximately 0.04, synthesised using 12.5 SCCM nitrogen flow rate. The different load versus displacement curves are shown in Figure 5.14(a)–5.14(e) and used to measure the hardness and Young's modulus values at these initial load conditions. The respective nanoindentation profiles are shown as an inset in the respective figures. The profiling of these indentations suggests that depth of such indentations increases with increasing the load with wider marked diameter. The measured *H* and *Y* values for this sample as a function of load, within the elastic limit, are summarised in Figure 5.14(f). These measured values suggest that initially both hardness and Young's modulus values are increasing with load up to approximately 140 μN and reached to nearly constant hardness and Young's modulus with any increase in load conditions.

5.6 THERMAL STABILITY IN AIR AND VACUUM OF ZrO_X/ZrC-ZrN/Zr ABSORBER-REFLECTOR TANDEM STRUCTURES

The thermal stability of these solar selective coatings is important especially for high-temperature applications maintaining high spectral selectivity. The STA measurements are shown in Figure 5.12(A and B) (b) suggesting that SSCs are thermally stable up to 700°C or higher in an inert atmosphere, as no peak has been observed in derivative of weight change (WC). In contrast to the inert ambient, these structures start degrading in air at much lower temperatures approximately 200°C (Figure 5.12(B) (b). This is mainly due to the oxidation of layered tandem structures, which is more plausible in the air, suggesting that the structure can be used for high-temperature application in a vacuum or inert environment whereas at moderate temperature (< 200°C) in air/open environmental conditions.

The sample S5 with optimised selectivity has been annealed at 700°C and 600°C in vacuum for one and half hours for SS and Cu substrates, respectively, and at 200°C for one hour in the air for both the substrates to understand the structure–property correlation and its impact on solar thermal performance. The details about the structural degradation and related vibrational properties of the annealed tandem absorber-reflector structure are explained by Usmani *et al.* [Usmani and Dixit, 2016a]. The microstructural measurements of heat-treated S5 coating surfaces on both SS and Cu substrates are shown in Figure 5.15a(i, ii, ii) and 5.15b(i, ii, ii), respectively. It can be seen that the degradation is clearly visible in these heat-treated samples with respect to as-deposited sample, as shown in Figure 5.15. The vacuum-annealed samples are less affected as compared to air-annealed samples. Moreover, the selective structure on stainless substrates is less affected with respect to that on copper substrates as shown in

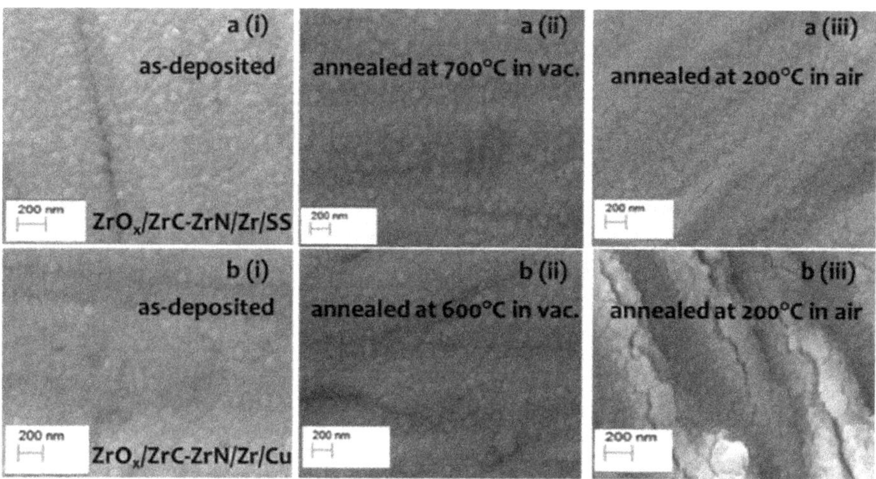

FIGURE 5.15 SEM micrographs of heat-treated ZrO_x/ZrC-ZrN/Zr absorber-reflector tandem structure for S5 sample on SS and Cu substrates: a(i) and b(i) as deposited; a(ii) and b(ii) annealed at 700°C/600°C in vacuum; and a(iii) and b(iii) annealed at 200°C in the air on SS and Cu substrates, respectively.

Figure 5.15. The large micro-cracks, in the case of air, annealed selective surface on copper substrates is because of the large thermal mismatch between the zirconium based layered structures, especially ZrO_x phase, and copper substrates. However, SS substrates show less thermal mismatch with zirconium-layered structure, especially ZrO_x phase, as shown in Figure 5.15a(i, ii, iii). In ambient air, large surface defects were generated due to the development of ZrO_x phase by transferring nitride and carbide phases in zirconium oxide phase. 3D AFM images of annealed S5 samples are shown in Figure 5.16, including vacuum and air-annealed structures on SS and Cu substrates. The roughness were increased by approximately 15% for the selective structures on SS substrates for both vacuum and air-annealed samples, whereas this became worse in case of selective structures on Cu substrates, and the roughness values have shown an increase up to approximately 47% and 82% with respect to as-prepared structures for 600°C vacuum and 200°C air-annealed samples. The microstructural and surface changes, especially in air-annealed case, may affect the optical response of the deposited tandem selective structures.

The reflectance spectra of these annealed tandem selective structures are shown in Figure 5.17(a) and 5.17(b) in 0.3–25 μm wavelength range, for S5 sample on SS and Cu substrates, respectively. This measured reflectance is used to calculate the absorptance and emittance in the desired wavelength range, and results are summarised in Figure 5.17(c) and 5.17(d), suggesting that the degradation is less significant for these structures on SS substrates as compared to that on Cu substrates,

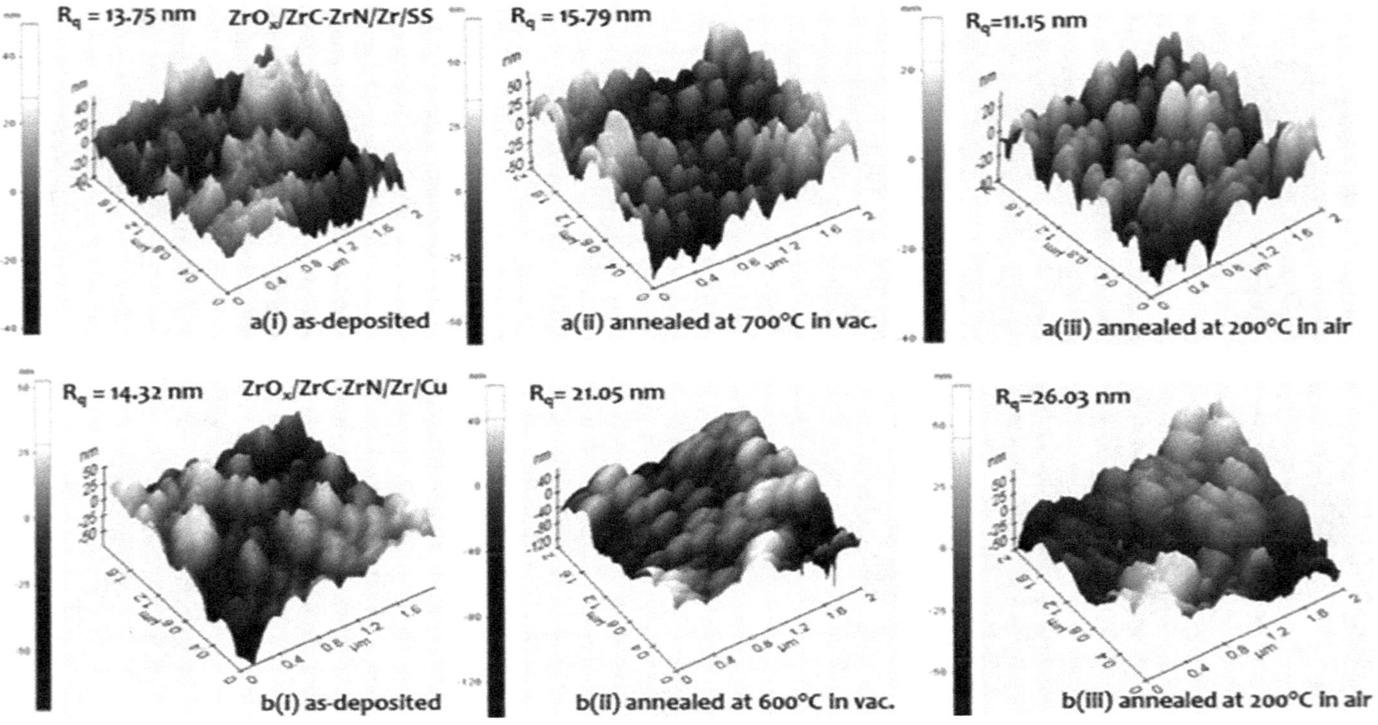

FIGURE 5.16 Three-dimensional AFM micrographs of ZrO_x/ZrC-ZrN/Zr absorber-reflector tandem structure for S5 sample on SS and Cu substrates: a(i) and b(i) as deposited; a(ii) and b(ii) annealed at 700°C/600°C in vacuum; and a(iii) and b(iii) annealed at 200°C in the air on SS and Cu substrates, respectively.

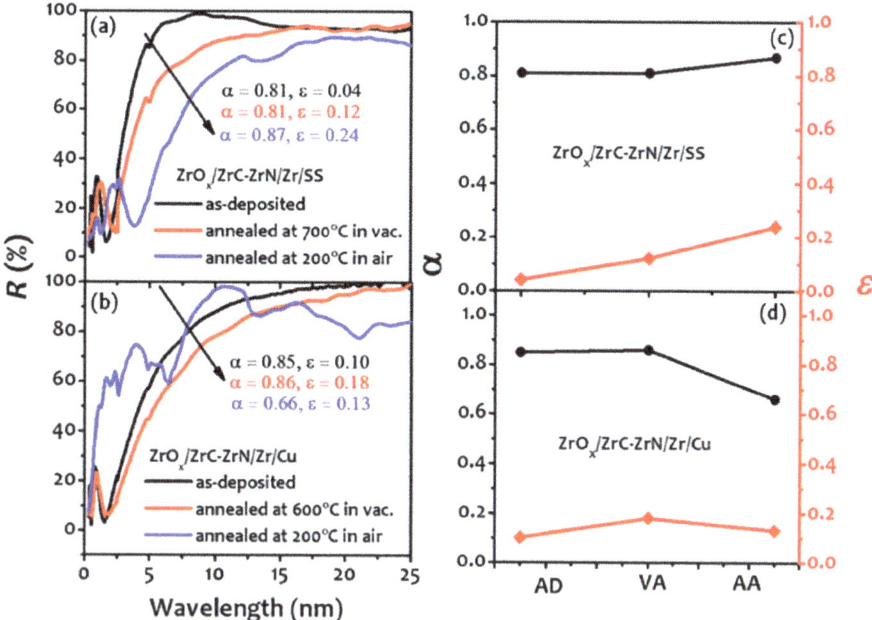

FIGURE 5.17 Percentage reflectance versus wavelength for ZrO_x/ZrC-ZrN/Zr absorber-reflector tandem structures: (a and b) as-deposited, heat-treated at 700°C/600°C in vacuum for 1.5 hours and at 200°C in air for 1 hour for S5 sample on SS and Cu substrates, respectively, and (c and d) absorptance (α) and emittance (ε) for as-deposited and heat-treated S5 sample on SS and Cu substrate, where acronyms AD, VA, and AA represent as-deposited, vacuum annealed and air-annealed samples, respectively.

where the absorptance has lowered approximately 23% after heat treatment. This is consistent with observed microstructural degradation in case of a tandem selective layer on Cu substrates.

5.7 CORROSION STUDIES ON MICROSTRUCTURE AND MECHANICAL PROPERTIES OF ZrO_X/ZrC-ZrN/Zr ABSORBER-REFLECTOR TANDEM STRUCTURES

In this section, the details of corrosion studies are discussed on absorber-reflector tandem selective structures and their correlation with solar thermal performance. Corrosion degradations are investigated in 3.5 wt.% NaCl (saline) solutions. The impact of these intended corrosion experiments has been investigated to understand the degradation process for pristine, vacuum, and air-annealed solar selective structures. The corrosion studies are investigated on ZrO_x/ZrC-ZrN/Zr structures deposited on both SS and Cu substrates, using electrochemical potential current density measurements. The details about the experimental process and related characterisations are explained by Usmani *et al.* [Usmani and Dixit, 2016b].

5.7.1 Results and Discussion

5.7.1.1 Corrosion Studies on Substrates and ZrO_x/ZrC-ZrN/Zr Structures in 3.5 wt.% NaCl Solution

The potentiodynamic polarisation measurements were carried out in 3.5 wt.% NaCl electrolyte solutions at different scan rates, as explained in Section 3.3.10.2. Bare SS and Cu substrates, as-deposited and thermally treated ZrO_x /ZrC-ZrN/Zr tandem structures on SS and Cu substrates, are used as working electrodes in three-electrode configurations. The measured voltages versus logarithmic of current results are summarised as Tafel plots in Figure 5.18 (a. These measurements were collected at different scan rates to understand their impact on corrosion dynamics. The estimated corrosion results are summarised in Table 5.4 (a and b) for SS substrate and tandem selective structures on SS substrate. The corrosion current (I_{corr}) and potential (E_{corr}) are taken at the intersection points of extrapolated anodic and cathodic Tafel curves of Figure 5.18. The polarisation resistance is calculated using the Stern–Geary equation $R_P = \frac{b_c \times b_a}{2.3 \times i_{corr}\left(b_a + b_c\right)}\left(K\Omega.cm^2\right)$, where b_c and b_a are the slopes of cathodic and anodic Tafel lines, i_{corr} is corrosion current density (μA/cm^2), and R_p is the polarisation resistance (KΩ. cm^2) [Stern and Geary, 1957, 37]. These results suggest that the corrosion current increases linearly with scan rates for all the samples under investigation, as indicated in the insets in Figure 5.18. The corrosion current (I_{corr}) is lower, approximately 0.555 μA, for tandem selective deposited SS substrate as compared to 2.07 μA for SS substrate at a maximum scan rate of 50 mVs^{-1}, suggesting that corrosion resistivity has increased for coated structures as compared to that of bare substrates. The similar behaviour has been observed for other scan rates, and respective corrosion currents are listed in Table 5.4 for these samples. Corrosion potential (E_{corr}) and polarisation resistance (R_p) values also increase with increasing scan rates. These values are lower for tandem selective structures on SS substrate with respect to that of bare substrates. The detailed measurements of respective parameters are summarised in Table 5.4 for SS substrate.

The scan rate has a strong effect on the corrosion parameters such as corrosion potential and corrosion current density. Figure 5.19(a) shows the respective changes in corrosion potential ($E_{corr.}$) and corrosion current density ($i_{corr.}$), as a function of scan rate. The corrosion potential of bare 304 SS is decreased from −0.1353 V to −0.4148 V as the scan rate increases from 1 mV/s to 50 mV/s (Figure 5.19a). Moreover, corrosion current density ($i_{corr.}$) increases from 0.0210 μA/cm^2 to 0.684 μA/cm^2 as scan rate increases from 1 mV/s to 50 mV/s (Figure 5.19(b)). This phenomenon occurs due to the fast attack of corrosion elements on the coating surfaces. However, at low scan rate (1 mV/s), corrosion potential of −0.0655 V of tandem structure deposited on stainless steel substrate increases from corrosion potential of −0.1353 V of bare SS substrate (Figure 5.19a). In contrast, the corrosion current density $i_{corr.}$, 0.00135 μA/cm^2, of tandem selective structure-coated SS substrate decreases as compared to that of the bare SS substrate, 0.0210 μA/cm^2 (Figure 5.19b). These measurements suggest that the selective surface-coated substrates have better corrosion resistance approximately 1,379 KΩ as compared to that of the bare substrate (382 KΩ) (Table 5.4). Thus, the observed increase in corrosion resistance for the tandem selective film on

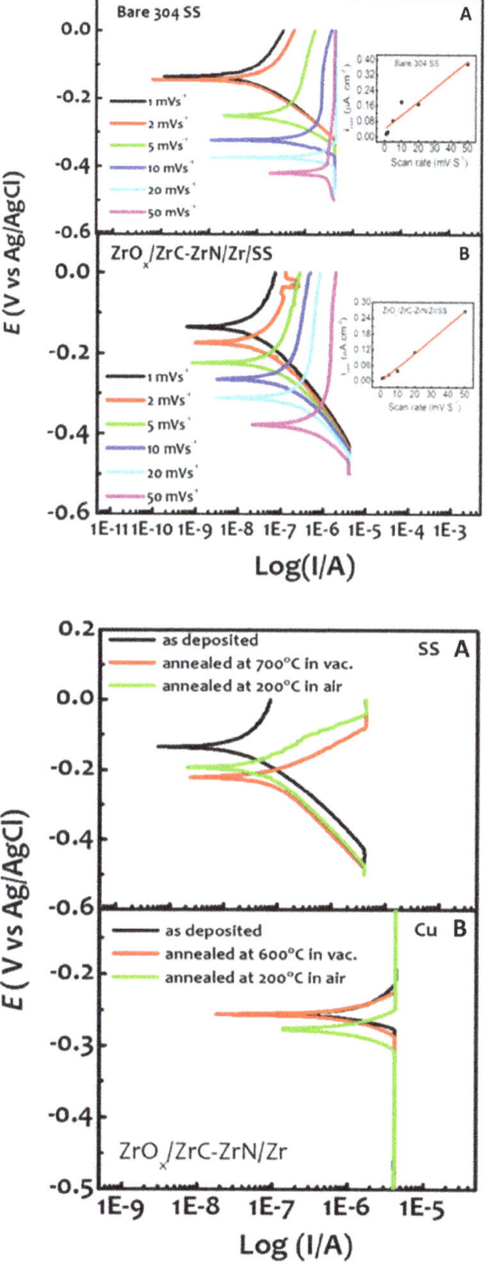

FIGURE 5.18 (a) Potentiodynamic polarisation curves for (A) bare 304 SS and (B) ARTSSCs (ZrO_x/ZrC-ZrN/Zr) deposited on SS substrates, in 3.5 wt.% NaCl solutions, for different scan rates. (b) Potentiodynamic polarisation curves of as-deposited, vacuum, and air-annealed ARTSSCs (ZrO_x/ZrC-ZrN/Zr) deposited on (A) 304 SS and (B) Cu substrates in 3.5 wt.% NaCl solution at scan rate 1 mVs^{-1}.

TABLE 5.4
The estimated corrosion parameter of (a) bare 304 SS substrate and (b) ZrO_x/ZrC-ZrN/Zr ARTSSCs deposited on SS substrate derived from Tafel analysis of polarisation curves.

(a) Tafel Data for 304 SS						
Scan rate	**1 mVs^{-1}**	**2 mVs^{-1}**	**5 mVs^{-1}**	**10 mVs^{-1}**	**20 mVs^{-1}**	**50 mVs^{-1}**
E_{corr} (V)	-0.1353	-0.1462	-0.256	-0.3276	-0.3752	-0.4148
I_{corr} (μA)	0.0635	0.07726	0.262	0.782	1.59	2.07
i_{corr} (μA.cm^{-2})	0.0210	0.0240	0.0866	0.258	0.526	0.684
R_p (K Ω)	382	286	72.1	23.1	10.6	18.7
b_a (V/dec)	0.159	0.107	0.163	0.164	0.103	0.181
b_c (V/dec)	0.086	0.087	0.059	0.056	0.063	0.175
C. Rate (mm/y)	0.000216	0.000247	0.000891	0.002655	0.005411	0.007032

(b) Tafel data for ZrO_x/ZrC-ZrN/Zr/SS						
Scan rate	**1 mVs^{-1}**	**2 mVs^{-1}**	**5 mVs^{-1}**	**10 mVs^{-1}**	**20 mVs^{-1}**	**50 mVs^{-1}**
E_{corr} (V)	-0.06555	-0.1025	-0.0973	-0.1018	-0.123	-0.1726
I_{corr} (μA)	0.02075	0.01923	0.0958	0.3116	0.637	0.555
i_{corr} (μA.cm^{-2})	0.00135	0.0042	0.014	0.0628	0.131	0.058
R_p (KΩ)	1379	769	470.95	257.45	134.7	46.65
b_a (V/dec)	0.098	0.11	0.122	0.153	0.146	0.195
b_c (V/dec)	0.067	0.065	0.062	0.0`73	0.078	0.091
C. Rate (mm/y)	0.000054	0.000164	0.0000745	0.000154	0.000364	0.00358

SS substrate is three times higher than that of the bare SS substrates. This is because the top zirconium oxide (ZrO_x) layer, which may act as a passivation layer to the absorber surface, may prevent the surface from corrosion attacks [Barshilia *et al.*, 2004; Grips *et al.*, 2006].

These observations suggest that tandem (ZrO_x/ZrC-ZrN/Zr) selective structures act as corrosion protecting layers in conjunction with their solar absorbing properties on SS substrate. The observed high corrosion resistance for spectrally selective ZrO_x/ZrC-ZrN/Zr structures is consistent with the reported literature [Liu *et al.*, 2001; Barshilia *et al.*, 2004; Shinn *et al.*, 1992; Aydoğdu and Aydinol, 2006; Huang *et al.*, 2007a, 2007b]. Zirconium oxide (ZrO_x) layer in the tandem selective structure also acts as a corrosion barrier against the bare substrate, because of its chemical and electrical inertness. The observed corrosion may be explained by introducing surface reactions as explained in Figure 5.20, where different ions from electrolytic solution start reacting with the surface and start penetrating into lower layered structures. The most common OH^-, Cl^-, and Na^+ electrolyte ions react with metallic Zr and nitrogen elements in the absorber layer and convert them into respective chlorides, nitrates, etc. The corrosion mostly initiates at surface micro-cracks or corners, where layered structures are relatively prone to such degradations.

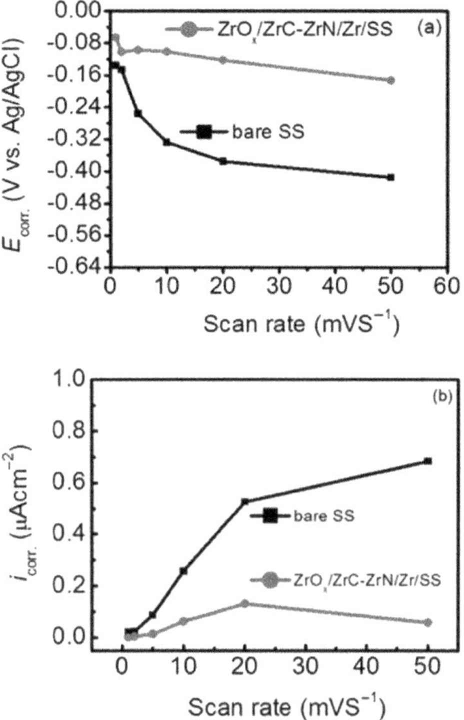

FIGURE 5.19 Variation of (a) corrosion potential ($E_{corr.}$) and (b) corrosion current density ($i_{corr.}$) of bare SS and ZrO_x/ZrC-ZrN/Zr absorber-reflector tandem structure coated on SS substrate with varying scan rates.

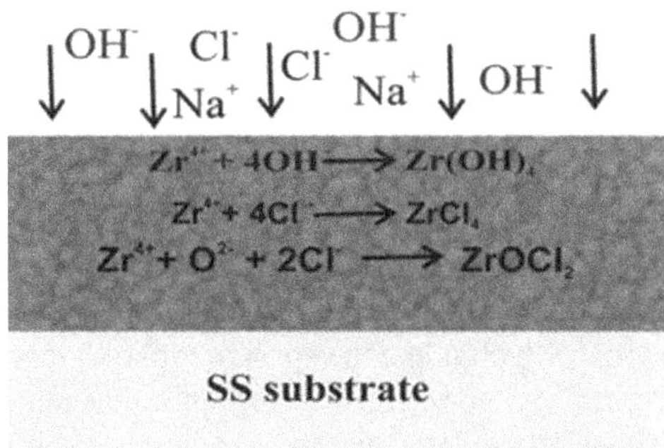

FIGURE 5.20 A schematic illustration of electrochemical corrosion attack of ZrO_x/ZrC-ZrN/Zr ARTSSCs on SS substrate in 3.5 wt.% NaCl solution.

The potentiodynamic polarisation curves of annealed tandem selective structure on SS and Cu substrates are shown in Figure 5.18 (b), at 1 mVs^{-1} scan rate in 3.5 wt.% NaCl electrolyte solution. The estimated corrosion parameters from potentiodynamic polarisation measurements are summarised in Table 5.5(a) and 5.5(b) for as-deposited (AD) and vacuum-annealed (VA) and air-annealed (AA), tandem selective structures

TABLE 5.5
The calculated values of corrosion parameters for as-deposited, vacuum, and air-annealed ZrO_x /ZrC-ZrN/Zr ARTSSCs deposited on (a) 304 SS and (b) Cu substrates, using potentiodynamic polarisation measurements.

(a) Tafel data for ZrO_x/ZrC-ZrN/Zr/SS at 1 mVs^{-1}			
	As-deposited (AD)	**Vacuum-Annealed (VA) (700°C)**	**Air-Annealed (AA) (200°C)**
E_{corr} (V)	−0.06555	−0.14475	−0.12355
I_{corr} (μA)	0.02075	0.03445	0.0027
i_{corr} (μA.cm^{-2})	0.00135	0.8605	0.0245
R_p (K Ω)	1379	104	435
b_a (V/dec)	0.098	0.066	0.063
b_c (V/dec)	0.067	0.143	0.068
C. Rate (mm/y)	0.000054	0.00144	0.000512

(b) Tafel data for ZrO_x/ZrC-ZrN/Zr/Cu at 1 mVs-1			
	As-deposited (AD)	**Vacuum-annealed (VA) (600°C)**	**Air-annealed (AA) (200°C)**
E_{corr} (V)	-0.1535	-0.1531	-0.1733
I_{corr} (μA)	0.303	0.192	0.253
i_{corr} (μA.cm^{-2})	0.722	0.574	0.386
R_p (K Ω)	7.223	9.974	8.548
b_a (V/dec)	0.048	0.038	0.032
b_c (V/dec)	0.022	0.03	0.028
C. Rate (mm/y)	0.01402	0.01138	0.00805

on SS and Cu substrates, respectively. These measurements suggest that the vacuum-annealed tandem selective structures on both SS and Cu substrates degrade significantly as the corrosion rates are nearly an order of magnitude higher as compared to the as-deposited structures. This is because of the induced oxygen vacancies during vacuum annealing, which make the surface prone to react at oxygen vacancy active sites in a corrosive environment, causing an enhancement in the corrosion rates. However, air-annealed tandem selective structures exhibit reduced corrosion rates, very similar to the pristine as-deposited structures. The air annealing of tandem selective structures does not create oxygen vacancies in the surfaces, leaving the surface intact or even enhanced oxygen stoichiometry unlike in the case of vacuum-annealed case. The stoichiometric ZrO_x top surface may assist in protecting structures from corrosion attacks, as confirmed by these experimental findings.

In addition, the polarisation measurements (for 1 mV/s scan rate) have been used for calculating the protective efficiency, P_i (%), for as-deposited (AD), vacuum (VA), and air-annealed (AA) samples using equation: $P_i(\%) = \left[1 - \left(\frac{i_{corr}}{i^0_{corr}}\right)\right] \times 100$, where i_{corr}

and i_{corr}^{0} represent the corrosion current density of tandem structure and substrate, respectively [Yoo *et al.*, 2008]. The AD selective structure on SS substrate showed the highest protective efficiency of 93.58% caused by lowest corrosion current density of 0.00135 μA/cm^{-2}, as compared to the VA and AA structures (Table 5.5). The porosity of these structures was calculated from corrosion potential (E_{corr}) and polarisation resistance (R_p) measurements (for 1 mV/s scan rate), inferred from the potentiodynamic polarisation technique, using equation $p = \left(\frac{R_{ps}}{R_p}\right) \times 10^{-\frac{\Delta E_{corr}}{b_a}}$, where P is the total coating porosity, R_{ps} is the polarisation resistance of the uncoated substrates, R_p is the polarisation resistance of the coated substrate, ΔE_{corr} the difference between the corrosion potential of the coated and uncoated substrate, and b_a is the anodic Tafel slope of the substrate [Alanyani and Souto, 2003]. The porosity of the as-deposited tandem selective structure coated on SS substrate is 9.93%, a lower value as compared to that of the air-annealed (7.40 × 10^{1} %), suggesting that the AD tandem selective structure had a low porosity as compared to the annealed one.

5.7.1.2 Structure–Property Correlation for Corrosion-Treated ZrO_x/ZrC-ZrN/Zr Structures

The structural and microstructural properties for these corrosion-treated tandem spectrally selective surfaces are explained by Usmani *et al.* [Usmani and Dixit, 2016b]. The corrosion resistance of these tandem selective structures also depends on the packing density of microstructure/granular geometries, and in general, the corrosion current density, i_{corr}, decreases with increasing the packing density. Corrosion-treated air-annealed tandem selective structures showed corrosion current density approximately 0.0245 μA/cm^2, which is less than that of vacuum heat-treated tandem selective structure, approximately 0.8605 μA/cm^2, but higher than as-deposited tandem selective structure, approximately 0.00135 μA/cm^2. Air-annealed selective structure showed a high corrosion resistance, suggesting enhanced stoichiometric ZrO_x formation as compared to that of heat-treated tandem selective structures under vacuum, which is oxygen deficient [Huang *et al.*, 2007a; Kubaschewski, 1993]. The oxygen-deficient ZrO_x is a relatively good electrical conductor, and oxygen vacancies provide corrosion-attacking sites and thus showing enhanced corrosion current density, consistent with experimental observations. Also, electrical conductivity is a good measure of the corrosion current density (CCD), and a higher electrical resistance may result in lower CCD i_{corr} at the same applied potential and thus may exhibit a higher corrosion resistance [Huang *et al.*, 2008]. The similar microstructural degradation and corrosion behaviour have been observed for corrosion-treated tandem selective structures on Cu substrates.

5.7.1.3 Mechanical Degradation of Corrosion Treated ZrO_x/ZrC-ZrN/Zr Structures

The measured load versus displacement curves are shown in the Figure 5.21 (a) for AD, VA, and AA tandem selective structures after corrosion treatment. The initial force was set at 14.3 μN for all nanoindentation measurements, and also measurements are corrected for any residual backgrounds. The nanoindented area A was

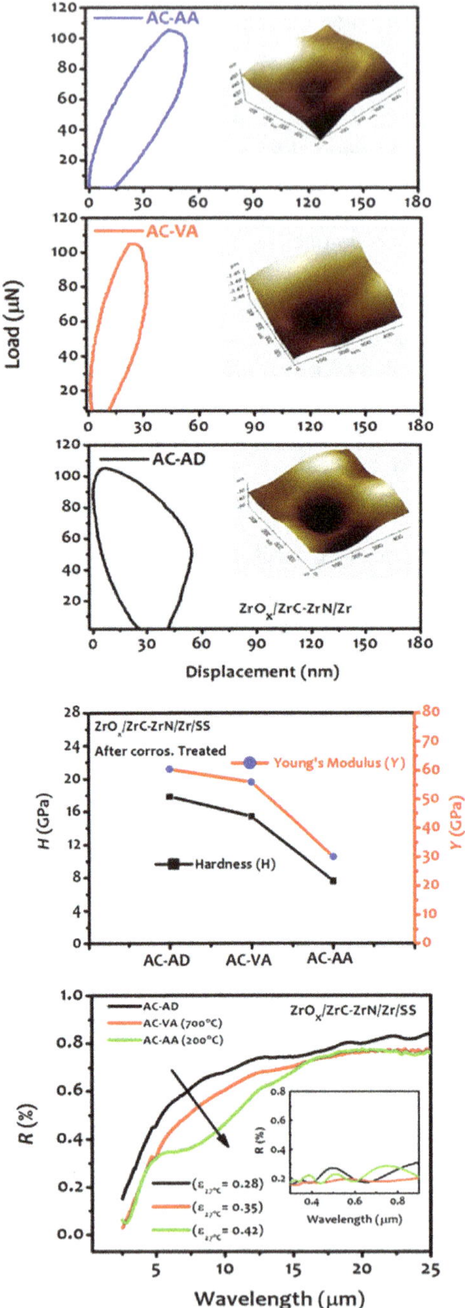

FIGURE 5.21 The load (P) versus displacement curves of corrosion-treated AD, AV, and AA ARTSSCs coated on SS substrate at load 120 μN, with inset showing representative AFM nanoindentation areal profile **(a);** Hardness (H) and Young's modulus (Y) of corrosion-treated AD, AV, and AA ARTSSCs coated on SS substrate at 120 μN **(b);** reflectance spectra of corrosion treated as-deposited (AC-AD), vacuum (AC-VA), and air (AC-AA) annealed ARTSSCs coated on SS substrates, in 3.5 wt.% NaCl solution **(c).**

estimated by averaging ten such indentations for calculating Young's modulus (YM) and hardness values for these structures, following Oliver and Pharr methodology for analysing the measurements [Oliver and Pharr, 1992, 2004].

The estimated YM and hardness values are shown in Figure 5.21 (b) suggesting that both reduced drastically for corrosion-treated air-annealed tandem selective structures. Also, both values reduced to approximately 50% of the untreated as-deposited and vacuum-annealed samples. This is because of relatively large microstructural degradation due to the oxygen diffusion inside multilayer structures for air-annealed samples. The observed reduction in both hardness and Young's modulus values for corrosion-treated air-annealed (AC-AA) tandem structure (from 17.85 GPa to 7.60 GPa and 60.48 GPa to 30.10 GPa) as compared to AD structures substantiates the observed relatively large microstructural degradation such as increased grain size and enhancement in the oxygen content in corrosion-treated air-annealed samples. The small amount of oxygen can modify the film properties, and oxygen segregation at the grain boundaries can also contribute to interface embrittlement [Riedl *et al.*, 2016]. This may lead to a change in fracture mode from intra- to inter-granular [De Hosson and Cavaleiro, 2006]. Moreover, increasing the content of oxygen in coating also acts as a softening factor in the coatings [Vaz *et al.*, 2004]. Hall–Petch relationship suggests that the materials' mechanical strength can be increased by reducing the grain size in a material [Sylwestrowicz and Hall, 1951; Petch, 1953]. Thus, a small grain size could be beneficial for mechanical strength. In addition, the oxygen impurities at grain boundaries in coatings may lead to the embrittlement because of the wrecking of the grain boundaries [Veprek and Veprek-Heijman, 2012]. Thus, the observed relatively poor values of YM and hardness are because of microstructural defects such as enhancement in grain size and oxygen content in the coatings, introduced during corrosion. The observed degradation has a severe impact on solar thermal properties of these structures.

5.7.1.4 Optical Degradation of Corrosion-Treated ZrO_x/ZrC-ZrN/Zr Structures

The tandem selective structures (ZrO_x /ZrC-ZrN/Zr/SS) exhibit an optical degradation after corrosion treatments. The observed degradations are consistent with the proposed corrosion attack pathways, as illustrated in Figure 5.20. To understand the impact on solar thermal performance, reflectance spectra were collected in the range of 0.3–0.9 μm (UV–Vis range) and 2.5–25 μm infrared range on these corrosion-treated tandem selective structures. The results are plotted in Figure 5.21 (c). The calculated absorptance and emittance values suggest that the absorptance of corrosion-treated tandem selective structure remained unaffected; however, emittance values increased drastically to 0.28, 0.35, and 0.42 for AD, VA, and AA tandem selective (ZrO_x /ZrC-ZrN/Zr/SS) structures. The observed increase in emittance is consistent with the observed structural and microstructural degradation, as emittance is a surface sensitive property. The corrosion has a severe impact on the microstructural degradation of surface structures for all investigated samples, which is directly related to the emittance values. The similar degradation of solar thermal properties has been observed for such spectrally selective structures on Cu substrates.

5.8 SOLAR PERFORMANCE ANALYSIS OF ZrO_X/ZrC-ZrN/Zr ABSORBER-REFLECTOR TANDEM STRUCTURES UNDER EXTREME THERMAL ENVIRONMENT

In this section, the effect of thermal degradation at elevated temperatures up to 900°C, in both nitrogen and air environments, has been discussed for ZrO_x /ZrC-ZrN/Zr structures. We carried out intensive structure–properties measurements after such extreme thermal treatments to understand the materials' degradation and related solar thermal performance.

5.8.1 Results and Discussion

5.8.1.1 Structural, Microstructural, Morphology, and Elemental Analysis

X-ray diffraction analysis on thermally treated 900°C structures at elevated temperature (up to 900°C) exhibited only tetragonal, monoclinic, and cubic phases of ZrO_x material. The SEM micrographs are shown in Figure 5.22 for both nitrogen and air-annealed ZrO_x /ZrC-ZrN/Zr/SS structures. The surface defects such as micro-cracks are clearly visible, as marked by arrows for clarity, in Figure 5.22, which are not seen in the case of AD samples, as discussed in previous sections. The AFM-measured surface roughness are much larger, approximately 26.21 nm and approximately 22.67 nm for heat-treated samples under nitrogen and air environments, respectively, as compared to that of approximately 13.75 nm for as-untreated ARTSSC samples [Usmani *et al.*, 2016d].

The observed degradation, in carbide and nitride phases of the absorber layer and their possible conversion into different zirconium oxide phase, as observed from XRD measurements, was also substantiated by EDX measurements on heat-treated ZrO_x /ZrC-ZrN/Zr/SS ARTSSC structures [Usmani *et al.*, 2016d]. A systematic

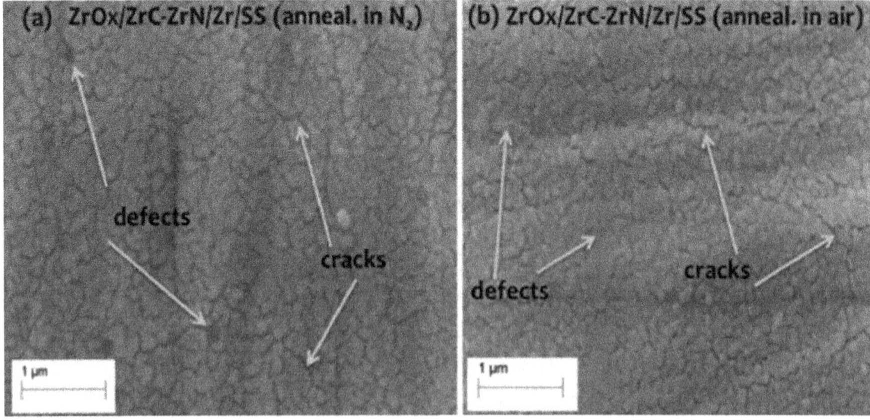

FIGURE 5.22 SEM images of heat-treated ZrO_x/ZrC-ZrN/Zr/SS ARTSSCs in (a) N_2 and (b) air environments up to 900°C.

increase in oxygen atomic fraction and decrease in nitrogen and carbon atomic fractions have been observed, and results substantiate the observed oxide phases only, as explained earlier. Increased atomic fraction of oxygen in these heat-treated ARTSSC structures, possibly, has increased defects and surface roughness as compared to the as-deposited structures. The absence of nitrogen content in these structures may be due to the EDX measurement limitations of the instrument and, thus, unable to notice the presence of any small nitrogen fraction in these samples, like the observed XRD measurements. The defect density and surface roughness values are much larger in the case of heat-treated samples under ambient air, with respect to that of samples treated under nitrogen environment. This is because of the enhanced oxygen atomic fraction, causing severe changes from carbide/nitride phases into ZrO_x phases, leading to the larger structural and surface damages in case of air-treated samples.

5.8.1.2 Mechanical and Optical Properties Analysis

The representative force–displacement curves, for 120 μN force, are plotted in Figure 5.5.23(A) with a respective AFM indentation profile (Figure 5.23(B)) on annealed tandem selective structure on SS substrate under air environment. Numerous such force–displacement curves were plotted for different force conditions and used to extract the variation of the hardness and Young's modulus as a function of applied force. The results are plotted in Figure 5.23(C) and 5.23(D) for annealed tandem selective structures under nitrogen and air environments, respectively. The hardness values decreased initially with force and saturated to values of approximately 7.07 GPa and approximately 5.42 GPa for annealed samples in nitrogen and ambient air. These values are much smaller than 18.88 GPa for AD tandem selective structure samples. This is consistent with our proposed hypothesis of ZrN-ZrC loss during heat treatments and also supported by XRD and EDX measurements. The similar behaviour has been observed on other spectrally selective structures [Huang *et al.*, 2007b, 2009, 2011a, 2011b; Fang and Chang, 2004]. The nitride and carbide phases are converted into oxides by the oxidation of ZrC-ZrN during heat treatment and, thus, resulting into reduced hardness for these structures. The ZrN phase is much harder than ZrO_x, and thus the loss of ZrN from the absorber structure may be attributed to the observed poor hardness values. In addition to these changes, the respective structural/microstructural degradation, such as cracks and larger gains of these heat-treated samples, is also responsible for the observed poor hardness values. In contrast to hardness, Young's modulus of these heat-treated samples did not exhibit a strong variation, as shown in Figure 5.23(D); yet, the measured values of Young's modulus for annealed sample in the air are always lower as compared to that of heat-treated in a nitrogen environment. This may be due to the relatively large microstructural defects of air-treated ARTSSCs with respect to that of nitrogen-treated ARTSSCs.

The observed structural/microstructural and surface changes (i.e. degradations) have an impact on optical response for these structures. The measured reflectance spectra for heat-treated ARTSSC structures are plotted in Figure 5.24 for both environmental conditions. We observed that the reflectance of heat-treated ARTSSC in the air is always lower as compared to that of in nitrogen-treated sample. This is again consistent with the observations of relatively larger defects in air-treated samples.

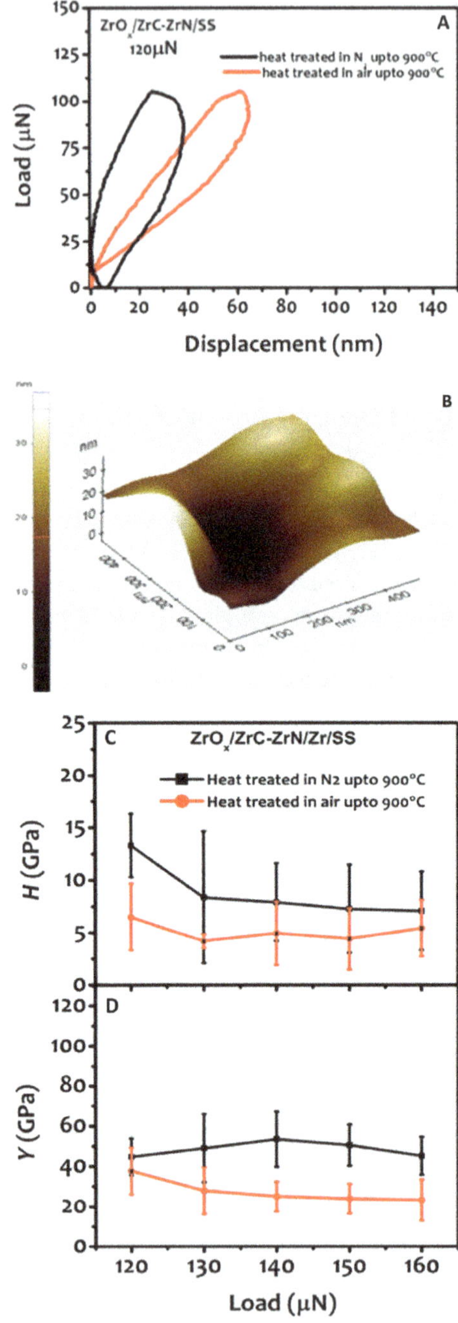

FIGURE 5.23 (A) Force versus displacement curve at 120 μN for heat-treated ZrO_x/ZrC-ZrN/Zr/SS ARTSSCs in N_2 and air environments up to 900°C, (B) AFM indentation profile of an indentation, (C) hardness (H), and (D) Young's Modulus (Y) versus applied force for heat-treated ZrO_x/ZrC-ZrN/Zr/SS ARTSSCs in N_2 and air environment.

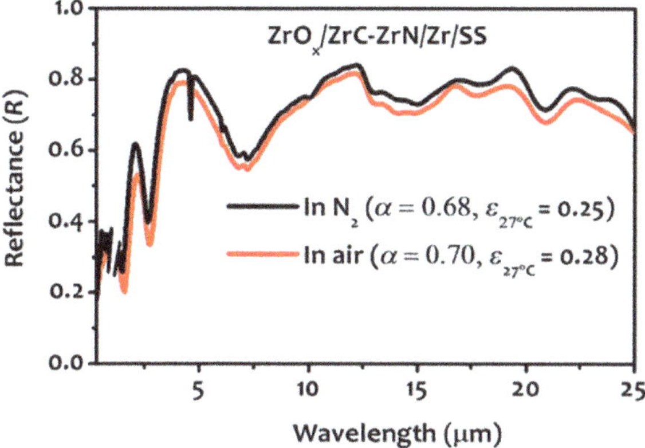

FIGURE 5.24 Reflectance spectra of heat-treated ZrO_x/ZrC-ZrN/Zr/SS ARTSSCs structures in N_2 and air environments up to 900°C.

The measured absorptance values are approximately 0.70 and approximately 0.68 for heat-treated ARTSSCs in nitrogen and air, respectively. These values are relatively smaller than that of as-deposited ARTSSCs (~ 0.88). The measured emissivity values are 0.25 and 0.28 for heat-treated ARTSSCs in nitrogen and air, respectively, and are much higher than that of as-deposited ARTSSC structure, approximately 0.04. The observed lower absorptance and higher emittance can be attributed to the enhanced structural and microstructural defects, developed during the thermal treatment of these ARTSSC structures under such extreme (~ 900°C) conditions.

6 Solar Selective Absorber Surfaces for Concentrating Solar Power Tower Technology

6.1 INTRODUCTION

Concentrating solar power (CSP) technologies are a potential candidate of renewable energy technologies. CSP technologies produce clean and efficient power with reducing greenhouse gas emission in comparison to conventional energy sources. Normally, CSP technologies are based on the concept of concentrating solar energy to produce steam or hot air which can then be used for electricity generation using conventional power cycles. Principally, CSP technologies are divided into two categories. First, the sun energy is concentrated via mirror arrangement on a line with a moderate solar radiation density up to 50 suns (e.g. Parabolic and Fresnel technologies). In the second case, the sun energy can be concentrated on a point with utilising heliostats up to 2,000 suns (e.g. Parabolic Dish and Solar Power Technologies) (shown in Figure 1.1(b) Chapter 1)). One of the capabilities of the CSP technologies is to store the solar energy during sun hours and to reutilise during night time, which makes the CSP technologies more attractive as compared to other renewable energy technologies. CSP technologies can also be hybridised with other renewable energy technologies such as photovoltaic and thermoelectric to increase the access of renewable energy power [Sarhaddi *et al.*, 2010; Yan and Malen, 2013]. The solar energy is absorbed by the receiver. In order to increase the absorptance of solar spectra, the receiver is coated with spectrally selective surfaces/films to absorb the maximum part of the solar radiation spectra. An ideal absorber must have high absorptance (~1) in the solar spectral region (0.3 μm to 2.5 μm) and lower thermal emittance (~0) in the infrared region (2.5 μm to 25 μm) (details are given in Chapter 2), stable in air, cost-effective, and should be easily fabricated [Aidroos *et al.*, 2015; Ambrosini *et al.*, 2017; Burlafinger *et al.*, 2015; López-Herraiz *et al.*, 2017; Merchán *et al.*, 2022; Nelson, 2020; Wu *et al.*, 2016].

Among the several CSP technologies, solar tower central receiver or solar power tower (SPT) technology is the most appropriate technology for achieving a high-temperature ($>1000°C$) process [Nelson, 2020; Wang *et al.*, 2023]. Figure 6.1 shows a schematic representation of SPT technology representing collector filed

DOI: 10.1201/9781003563990-6

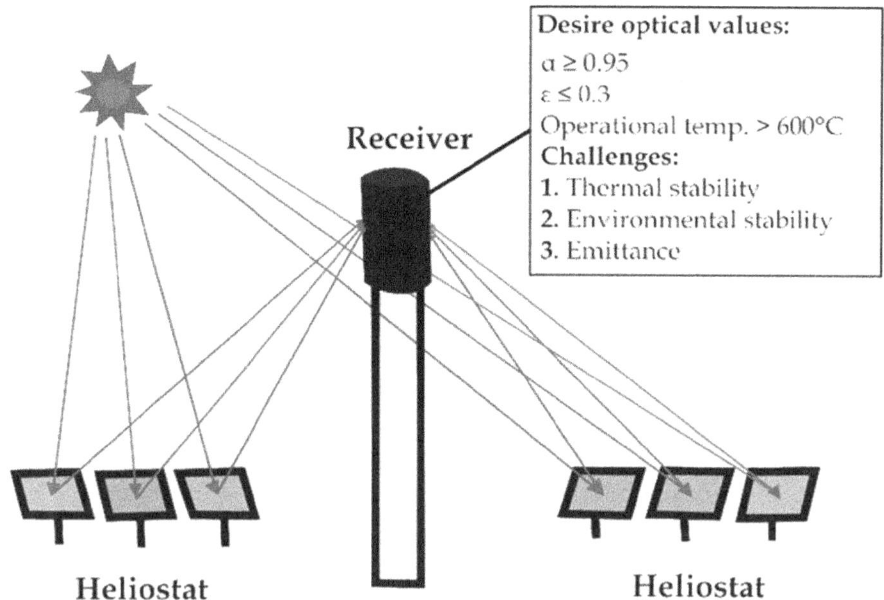

FIGURE 6.1 Schematics of solar tower technology, highlighting absorbers.

which is composed of heliostats and a receiver mounted on the top of the tower depicted with desired optical values. Heliostats of the SPT system track the sun and focus the solar radiation on the top of the tower. By concentrating solar radiation approximately 100 times, the receiver achieves temperature more than 1,000°C. The solar energy is absorbed by working fluid and used to generate steam to run a conventional turbine. In this chapter, we review the development in the solar selective absorber surfaces (SSASs) by solar power tower (SPT) technology (*T* more than 1,000°C) and discuss their analysing methods.

6.2 SOLAR SELECTIVE ABSORBER SURFACES FOR SOLAR POWER TOWER

Receiver is one of the important parts of the solar power tower technology and plays a crucial role in the thermal performance of the STP system. The receiver geometry, solar selective absorber materials' properties, and heat transfer fluid (HTF) are the three main parameters which can affect the performance of receiver and overall efficiency of the SPT technology. To improve the efficiency of the receiver, it is required to study the optimisation of geometry and solar selective absorber materials and increase the heat transfer between the receiver walls and the HTF [Larrouturou *et al.*, 2016]. SSASs coated on receiver of the SPT technology play a crucial role to absorb solar radiation and increase the temperature of the HTF above 1,000°C. Even if SSAS multilayers and graded cermets/tandem structures have been developed and successfully operated in vacuum in parabolic trough CSP technology [Cao *et al.*, 2014b; Céspedes *et al.*, 2014; Selvakumar and Barshilia, 2012], however, none of the SSASs structures have been designed and developed, which can be operated at higher temperature ($> 1000°C$) in air, and this is the requirement of SPT technology. Here, we review the so-far developed SSAS materials and structure, especially by SPT technologies. These are described in the next sections.

6.2.1 Pyromark 2500®

Pyromark 2500® is a commercially available black paint. It is a high-temperature-based silicon paint, which is manufactured by Tempil. It has a high solar absorptance ($\alpha > 0.95$, T = 600° C) as prepared on steel, aluminium, alloys, and ceramics substrate [Ambrosini *et al.*, 2019; Ho *et al.*, 2016]. Presently, Pyromark 2500® is used as the SSAS material in SPT technology. However, it faces large thermal losses with thermal emittance ($\varepsilon = 0.86$), during high-temperature operation. It is also degraded over a period of time, as operating at high temperature in air causes degradation in overall performance of an SPT system [Hall, 2012]. However, the absorptivity of Pyromark 2500® degrades after a high-temperature ($> 700°C$) exposure due to crystal structure changes and phase instability [Ho *et al.*, 2016].

6.2.2 Black Oxide (Co_3O_4)

For several decades, cobalt oxide materials have been studied as solar selective absorber surfaces. However, these show that the temperature stability range of cobalt oxide is from 300 to 650°C [Aeronautics, 1980; Film *et al.*, 1982; Cathro, 1984]. Moreover, Moon *et al.* reported that SSASs based on cobalt oxide nanoparticles exhibited unprecedented a high-temperature durability, showing no degradation in structural and optical properties after annealing at 750°C in air for 1,000 h [Moon *et al.*, 2015].

6.2.3 Iron Oxide (Fe_2O_3)

Wu *et al.* studied a low-cost method to grow Fe_2O_3 film using thermal oxidation on stainless steel (SS 304) substrate for obtaining a high-temperature solar absorber. Iron oxide films on SS 304 substrate showed a high absorptivity (α) (0.99–0.922) at a temperature range of 900 to 1,000°C, and thermal emittance (ε) values being relatively low from 0.18 to 0.38 [Wu *et al.*, 2016].

6.2.4 Metal Oxide Spinel

Commonly, spinel oxide has AB_2O_4 structures arranged in a cubic close-packed lattice, where A represents a divalent cation that occupies the tetrahedral space and B represents a trivalent cation that occupies the octahedral space [Rubin *et al.*, 2019].

Rubin *et al.* systematically studied the optical properties and long-term thermal stability of SSASs made from various Cu(II)-containing spinel oxide nanoparticles, including $CuCr_2O_4$, $Cu_{0.5}Cr_{1.1}Mn_{1.4}O_4$, and $CuFeMnO_4$, and compared these properties to those of the state-of-the-art Pyromark 2500 coating. They found that the porous $Cu_{0.5}Cr_{1.1}Mn_{1.4}O_4$ had the highest solar absorptance at 97.1% before the thermal annealing and remained the highest throughout thermal testing, remaining at 97.2% after 2000 h, whereas Pyromark 2500 exhibited considerable degradation in solar absorptance, from 25 % to 94.6% [Rubin *et al.*, 2019].

Karas *et al.* studied the bi-metallic nanoparticles of copper-cobalt oxide ($Cu_{0.15}Co_{2.84}O_4$) and copper-manganese oxide ($Cu_{1.5}Mn_{1.5}O_4$) spinel types for high-temperature solar absorber-coating applications. The material is deposited onto high-temperature, durable Inconel substrates by a flexible spray-coating method, and characterisation is performed by Scanning Electron Microscopy (SEM), Energy-Dispersive X-Ray Spectroscopy (EDS), and X-Ray Powder Diffraction (XRD) analyses, as well as measurements to gauge thermal performance were taken. High-temperature stability of a model solar receiver surface using these synthesised materials is assessed by comparing spectral reflectance and a figure-of-merit efficiency metric before and after high-temperature exposure beyond 1,000 h. This study ultimately showcases materials produced with high figure-of-merit conversion efficiency, demonstrating solar absorber coatings capable of interfacing with next-generation CSP receiver systems [Karas *et al.*, 2018].

Wang *et al.* designed and demonstrated air-stable, $MnFe_2O_4$ nanoparticle-pigmented solar selective coatings with a high solar absorptance of approximately 93%, a relatively low thermal emittance of approximately 52%, and an optical-to-thermal energy conversion efficiency of 89.7% under 1,000× solar concentration at 750°C. Guided by four-flux radiation model, high spectral selectivity is achieved using cost-effective spray-coating approach, a notable improvement over conventional vacuum-deposited, multilayer solar selective coatings for low-cost, high-efficiency CSP receivers. The thermal degradation of benchmark Pyromark 2500 coatings happens at 750°C after 300 h, the solar absorptance of the $MnFe_2O_4$-pigmented coatings on stainless steel 310 substrates is increased to approximately 93%, and the optical-to-thermal energy conversion efficiency is improved to 89.7% after serving at 25°C in air for 700 h. This performance enhancement is attributed to the transformation of $MnFe_2O_4$ NPs into more thermodynamically stable, non-stoichiometric manganese-rich manganese ferrite and iron-rich manganese iron oxide phases, driven by an increase in configurational entropy that favours Gibbs free energy reduction at high temperatures. For more than 1,000 hour-endurance testing at 750°C in air plus 19 day–night thermal cycle testing between 750°C (12 h/cycle) and 25 °C (12 h/cycle) on SS 310 substrates, the thermal degradation is mainly due to the CrO_x micro-flake formation from SS 310 substrates rather than the coatings, which can be further suppressed by pre-oxidising the surface of SS 310. With lower emittance matrix material and further optimisation of pigment nanoparticle composition, stoichiometry, concentration, and coating thickness, it is promising to achieve an optimised thermal efficiency of 92.5% with a long-term thermal stability at 750°C for Generation 3 CSP systems [Wang *et al.*, 2021].

Xu *et al.* reported the spray-coated spinel Cu-Mn-Cr oxide nanoparticle-pigmented solar selective coatings on Inconel tube sections maintaining $\geq 94\%$ efficiency at 750°C and $\geq 92.5\%$ at 800°C under 1000x solar concentration. The solar spectral selectivity is intrinsic to the band-to-band and d-d transitions of non-stoichiometric spinel Cu-Mn-Cr oxide nanoparticles by balancing the lattice site inversion of Cu^{2+} and Mn^{3+} on tetrahedral versus octahedral sites. This feature offers a large fabrication tolerance in nanoparticle volume fraction and coating thickness, facilitating low-cost and scalable spray-coated high-efficiency solar selective absorbers for high-temperature CSP systems [Xu *et al.*, 2022].

6.2.5 Miscellaneous

Gray *et al.* designed and developed the metal silicide binary compound of refractory metals (TiSi, TaSi) as multilayer, thin-film stack solar selective absorber surfaces. Low emittance and other properties of these materials indicate that the SSASs of these materials should be oxidation-resistant in the air up to the temperature range 800°C–1200°C. Spectral measurements show that these coatings continue to perform at targeted levels after cycling to temperatures of 1000°C in environments of nitrogen and forming gas [Gray *et al.*, 2015]. Ambrosini *et al.* also explored the lanthanum Strontium Maganite (LaSrMnO) (LSM) as solar selective absorber surfaces for solar power tower technology [Ambrosini *et al.*, 2015]. Hall *et al.* prepared the Ni-25 graphite, Ni-5Al, and WC-20Co SSASs through thermal spray-coating techniques and $NiCo_2O_4$, $CuCo_2O_4$, and $(NiFe)CoO_5$ spinel SSASs and claimed that these are optically competitive with Pyromark paint [Hall, 2012].

6.3 ANALYSIS OF SOLAR SELECTIVE ABSORBER SURFACES OF SOLAR POWER TOWER

The parameters, solar absorptance (α) and thermal emittance (ε), define the spectral selectivity of the solar selective absorber surfaces. The details about the analysis of these parameters are explained in Chapter 2 (Section 2.2). The solar absorptance of a material is calculated from the direct reflectance measurement. A spectrophotometer with an integrating sphere attachment is used for collecting the reflected specular and diffuse components of the material to obtain the total reflectance versus wavelength values. The reflectance measurements are made over the portion of the electromagnetic spectrum from 0.3 to 2.5 mm, because approximately 95% of the solar radiation is covered by this range of spectra. The importance of the solar absorptance for thermosolar receivers is that the absorbed solar radiation is the predominant external heat input to the receiver. The relevance of thermal emittance is that it controls the rate at which heat leaves the tubes of the receiver [Fang *et al.*, 2014; López-herraiz *et al.*, 2017].

Solar Selective Absorber Surfaces have also been analysed on the basis of different temperatures. At high temperatures, radiative losses take more importance, that is the decrease of emittance plays an important role in impacting the performance of a receiver. This effect has been studied through the Figure of Merit (FOM). The absorber efficiency (η_{abs}) or FOM is defined as the net radiative energy absorbed

(absorbed incident solar radiation minus thermal radiative loss) divided by the total incident solar radiation:

$$\eta_{abs} or(FOM) = \frac{\alpha_{sol} Q - \varepsilon \sigma T^4}{Q} = \alpha_{sol} - \frac{\sigma T^4}{Q} \varepsilon$$

Where α_{sol} is the solar absorptivity, Q is the irradiance on the receiver (W/m^2), ε is the thermal emittance, σ is the Stefan–Boltzmann constant (5.67×10^{-8} W/m^2/K^4), and T is the surface temperature (K).

The FOM reflects the idea that maximising absorptance at the central receiver does more to improve receiver efficiency than minimising thermal emittance. The magnitude of energy absorbed by the receiver depends directly on the energy flux magnitude incident upon the surface. However, the magnitude of energy emitted by the receiver is only affected by the incident radiation if this flux leads to an increase in the receiver body temperature. Besides that, the SSASs are opaque to solar energy; therefore, maximising the receiver absorptance minimises the reflectance from the receiver surface [Hall, 2012; Ho, 2017; López-herraiz *et al.*, 2017].

Analysing method of measuring the mechanical properties such as Young's Modulus (YM) and hardness of solar selective absorber surfaces is explained in detail in Chapter 3 (Section 3.3.9), and analysing method of measuring the environmental effect (corrosion analysis) of SSASs is detailed explained in detail in Chapter 3 (Section 3.3.10).

6.4 CONCLUDING REMARKS

The efficiency of concentrating solar power tower technology can be improved by increasing the receiver temperatures. However, high temperature of receiver can modify the properties solar selective absorber surfaces. SSASs are durable and low cost and have high absorptivity and low emissivity for this application. Presently, Pyromark paint is the standard SSAS for a receiver of solar power tower technology. In this chapter, we reviewed the progress in SSASs by STP technology and summarised the analysing processes such as absorptance, emittance, FOM, mechanical properties, and environmental effect of SSASs of SPT technology.

7 Conclusion and Scope of the Future Work

Spectrally selective absorber surface is one of the most important components of the solar thermal energy technologies/systems. The higher operating conditions of receivers, consisting of absorber coatings, can enhance the solar thermal efficiency and reliability of the systems. Considering the same, different types of spectrally selective absorber surfaces have been modified/developed using different synthesis techniques. These include:

1. The modification of black chrome spectrally selective structures with graphite-encapsulated FeCo nanoparticles for mid-temperature solar thermal applications. The important developments are:
   1.1 The modified FeCo(C) NP black chrome structures are thermally stable up to approximately 400°C or more in open ambient conditions.
   1.2 The modified FeCo(C) NP black chrome structures exhibit relatively a larger corrosion resistance approximately 0.1968 KΩ as compared to pristine black chrome selective surface approximately 0.06684 KΩ (on the copper substrate) and thus can be used under extreme environmental conditions.
   1.3 Using 0.1 wt.% FeCo(C) NPs modified black chrome selective coatings, corrosion rate reduces to half (6.564 mm/y), as compared to the pristine black chrome (11.08 mm/y) selective coatings.
2. The design and development of absorber-reflector tandem (ZrO_x /ZrC-ZrN/Zr) selective structures for high-temperature applications. Some of the important developments are:
   2.1 Optimisation of metallic zirconium (Zr) layer for minimum thermal emittance (~ 0.12) in the desired wavelength range (2.5–25 μm). The optimised deposition condition includes DC sputtering of Zr infrared reflector at 350°C deposition temperature for two hours.
   2.2 Optimisation of ZrC-ZrN absorber layer using RF sputtering at different nitrogen flow rates during absorber deposition. The optimised absorber layer showed an absorptance of approximately 0.88 and the corresponding emittance is approximately 0.04 for ZrO_x /ZrC-ZrN/Zr structures on SS substrates.
   2.3 The mechanical properties such as Young's modulus and hardness of tandem absorber structure (ZrO_x /ZrC-ZrN/Zr) showed a strong dependence on growth conditions of the absorber layers and found that moderate Young's modulus and hardness are important for an enhanced solar thermal performance.

DOI: 10.1201/9781003563990-7

2.4 These ZrO_x /ZrC-ZrN/Zr structures are thermally stable up to approximately 700°C on SS and 600°C on Cu substrates in vacuum, and up to 200°C in air.

The impact of corrosion on this absorber-reflector tandem structure (ZrO_x /ZrC-ZrN/Zr) in 3.5 wt.% NaCl saline environments has been investigated. It is observed that ZrO_x /ZrC-ZrN/Zr structures on SS or Cu substrates exhibit larger corrosion resistance of approximately 1379 kΩ, as compared to bare stainless steel and copper substrates. Thus,ZrO_x /ZrC-ZrN/Zr structures can be used under adverse environmental conditions with an enhanced thermal and corrosion reliability.

The developed spectrally selective coating structures during these studies may further be subject to the development of thermally stable structures on specific substrates such as cylindrical pipes or other curved surfaces for realising the probable applications in real systems. The work can be extended to develop:

1. The low-cost deposition techniques such as electrochemical deposition, spray-coating technique, dip-coating, and sol-gel techniques for the realisation of large-scale spectrally selective absorber surfaces.
2. FeCo(C) NPs modified black chrome selective coatings can be developed for large-scale absorber tube employing electrodeposition techniques for high-temperature applications in inert or open ambient air.
3. The absorptance of the zirconium carbonitride (ZrC-ZrN) absorber layer can further be enhanced with varying absorber-layered structures, and oxide-based absorbers may be developed for an enhanced thermal and corrosion stability.
4. The absorber-reflector (ZrO_x /ZrC-ZrN/Zr) tandem structure can be developed by a large-scale coating employing sputtering system.

. . .

Annexure A: Miscellaneous Work

In this chapter, the work on ZnO nanorod-based electrodes has been discussed for hydrogen evolution and storage. These studies suggest the possibility of ZnO nanorods as hydrogen storage media in alkaline conditions.

A.1 ZnO NANOROD ELECTRODES FOR HYDROGEN EVOLUTION AND STORAGE

The molecular hydrogen (H_2) has been investigated intensively as an alternative to the fossil fuels because of its high specific heat of combustion. The hydrogen fuel cell may offer particular advantages for transportation because of its reduced weight, superior energy conversion efficiency, and ecological friendliness [Veziroglu and Barbir, 1995]. Energy-efficient hydrogen production and safe storage remain the key technological obstacle for H_2 based economy. The steam reforming of hydrocarbons is used for large-scale H_2 production. However, this process is not economical, and also the produced hydrogen is not pure enough to be used directly for an application. In contrast, the electrolysis of water has been used to produce relatively pure and economical hydrogen which can be used directly in fuel cells. Also, this hydrogen is free from carbon monoxide, usually present in steam reforming process, and is detrimental to the proton exchange membrane of fuel cells [Rossmeisl *et al.*, 2007]. In the hydrogen evolution reaction (HER) [Roberge, 2008], iron (Fe) and Ni (nickel) electrodes are used because of their relatively high H_2 evolution overpotentials (380 and 480 mV, respectively). In spite of high overpotentials, the process is energy intensive and consumes huge electricity [Shan *et al.*, 2008; De Giz *et al.*, 1995; Wang *et al.*, 2005; Arul, 2000; Qing *et al.*, 2003, 2004; de Souza Roberto *et al.*, 2007; Joanna and Antoni, 2007]. In recent years, numerous studies have focused on the quest for new electrode materials presenting a higher electrocatalytic efficiency for HER [Rosalbino *et al.*, 2005; Metikos-Hukovic *et al.*, 2006; Hu, 2000]. Many semiconductor oxides, such as TiO_2, WO_3, ZnO, $SrTiO_3$, Fe_2O_3, Cu_2O, and SiO_2, have been tested (in both bulk and nano forms) for such applications [Satsangia Vibha *et al.*, 2008; Bjorksten *et al.*, 1994; Zaban *et al.*, 2000; Aroutiounian *et al.*, 2005; Rau Greg, 2004; El-Meligi and Ismail, 2009]. There are few studies on the oxidation activities on ZnO as well as on the ability of ZnO nanorods (NRs) to store the hydrogen [Cox *et al.*, 2001a, 2001b; Hofmann *et al.*, 2002; Kılıç and Zunger, 2002; Lee *et al.*, 2006; Van de Walle, 2000, 2001, 2002]. ZnO has also been investigated for its quality of being used as photocatalytic anodes in solar-powered photo-electro-chemical (PEC) electrolysis, where H_2 is generated directly by using a metal cathode and a semiconducting anode for water splitting [Yan *et al.*, 2007; Wolcott *et al.*, 2009]. In this chapter, we will discuss the HER electrocatalytic efficiency and the hydrogen storage capability of high-quality PLD-grown ZnO NRs on Si (100) substrates (ZnO NR/Si (100)) in both acid and alkaline electrolytes.

A.1.1 Experimental Details

Vertical arrays of self-forming, catalyst-free, ZnO NRs were grown on Si (100) substrates using Pulsed Laser Deposition (PLD) by Rogers *et al.* and reported in details elsewhere [Rogers *et al.*, 2011]. The crystalline quality of ZnO nanostructures was investigated using X-Ray Diffraction (XRD) measurements, performed in a Panalytical MRD Pro system using a Cu K_α source, a four-bounce Ge monochromator in the incident beam path, and a "triple-axis" monochromator in the diffracted beam path. Sample morphology was studied using a Hitachi S4800 Field Emission-Scanning Electron Microscope (FE-SEM). The room-temperature Photo Luminescence (PL) measurement was carried out using 325-nm HeCd laser emission line. Electrochemical studies were done using ZnO (NR)/Si (100) as the cathode (working electrode), a platinum wire as the anode, and Ag/AgCl as a reference electrode. Electrical connections were made to the ZnO NR/Si (100) by soldering a copper wire to the Si. The fabricated electrochemical cells were tested using both 1 M H_2SO_4 acidic and 8.5 M KOH alkaline electrolytes. All cyclic voltammetry measurements were conducted using an Iviumstat spectro-electro-chemical workstation in the three-electrode assembly (details are discussed in Section 3.3.10.1).

A.1.2 Results and Discussion

The X-ray diffraction measurements were collected in 2θ/Ω and Ω scans, and results are shown in Figure A.1. The broadened multi-peak diffraction at 2θ which is approximately 34.47° suggests that the lattice parameter 'c' has increased from approximately 5.192 Å to approximately 5.200 Å. There is also some higher-angle broadening at the base of the peak, as shown in Figure A.1a, suggesting the contribution from smaller c lattice parameters as well. The disorders in these nanostructures, especially for the less dense a–b plane at the base of the NRs (and thus a larger lattice parameter 'a'), may be the main reason for the observed large variation in the lattice parameters [Sandana *et al.*, 2011]. The Ω scan measurement is also shown in Figure A.1b which is symmetrical with a Full-Width Half Maximum (FWHM) of 0.47°. These measurements explain that NRs are highly c-axis oriented.

Figure A.2(a) shows an FE-SEM image of the top surface of ZnO NRs on Si substrate. These rods are vertically aligned with diameters ranging from 100 nm to 300 nm, and a pitch of about 400 nm. The rods lengths were varied in between approximately 1 μm to approximately 2 μm by controlling the growth parameters. The room-temperature photoluminescence spectrum for the ZnO NRs on Si (100) substrate is shown in Figure A.2(b). Strong main emission peaked at 3.176 eV, corresponding to the wurtzite ZnO near band edge emission (NBE), with FWHM about 110 meV. The additional peak at 3.3 eV has been observed as shoulder peak and may correspond to the oxygen defects, especially oxygen vacancies consistent with the reported literature [Tam *et al.*, 2006; Wu and Liu, 2002].

The cyclic voltammetry plots for ZnO NRs/Si (100) in 1 M H_2SO_4 and 8.5 M KOH are shown in Figure A.3 (a and b). These observations suggest that H_2 evolution is irreversible in acidic, whereas reversible in basic conditions.

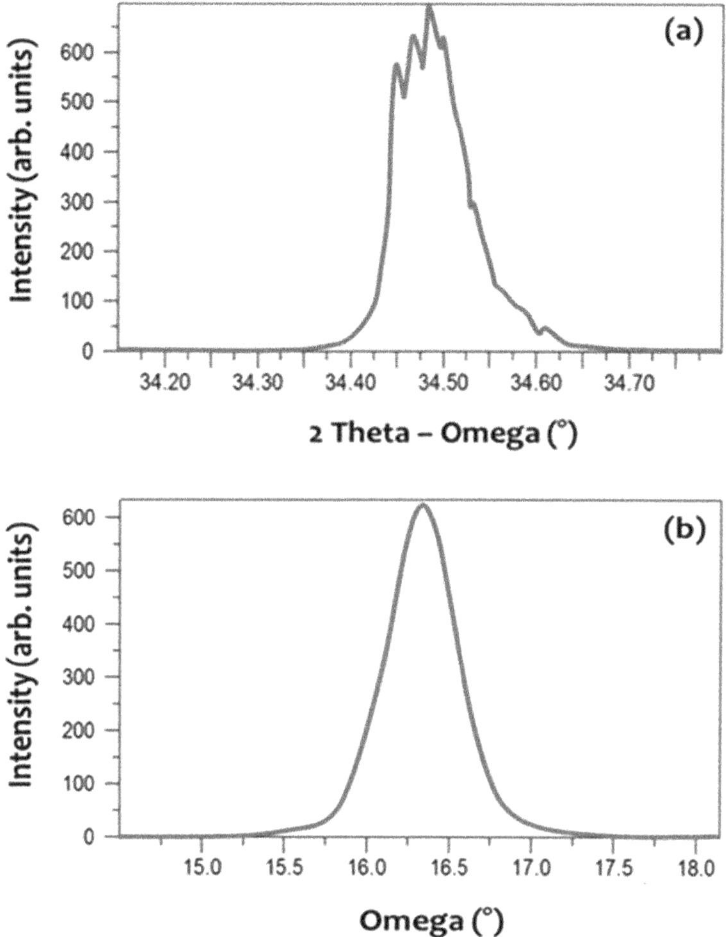

FIGURE A.1 XRD (a) 2θ/Ω and (b) Ω scans for the (0002) peak of the ZnO NRs grown on Si (100).

Figure A.3 (a) shows the irreversible nature of the voltammogram. To understand the hydrogen evolution reaction (HER) in an acidic electrolyte, three reaction mechanisms are considered, such as Volmer (electrosorption), Heyrovsky (electrodesorption), and Tafel (recombination) mechanisms [Harrington and Conway, 1987]. The electron transfer process for each mechanism is summarised in Table A.1 for acidic medium. Considering the similar possibility at the surface of ZnO NRs (Table A.1), hydrogen may form and escape through the solution, and that is why there is no reversible process in these voltammograms. It could also be possible that the generated H_2 may get trapped in the ZnO NRs due to Van der Waal's forces. The slight agitation would then be likely to release this physisorbed H_2.

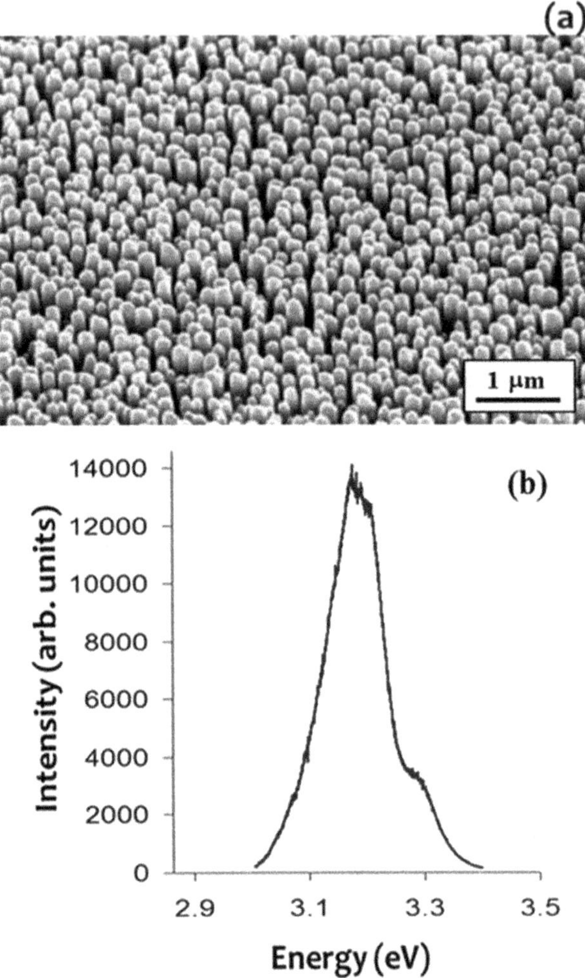

FIGURE A.2 (a) FE-SEM image of vertical ZnO nanorod array grown by PLD on a Si (100) substrate. (b) Room-temperature photoluminescence spectrum for the ZnO NR grown on Si (100).

Hydrogen evolution mechanism in the alkaline medium on both metal surface and ZnO surface is given in Table A.2 for the three possible mechanisms. The hydrogen evolution reaction is reversible and reproducible under alkaline conditions. The measured reversible cyclic voltammograms are shown in Figure A.3b.

If out of these three steps, Heyrovsky electrochemical desorption step becomes the rate-determining step (RDS) (i.e. evolution of H_2 from ZnO surface is the RDS), the resident time that an H atom is hydrogen-bonded to the O of ZnO will increase, and hence the cathodic peak will be smaller and broader. During the anodic cycle, the peak current increases, and a shift in peak potential between cathodic and anodic

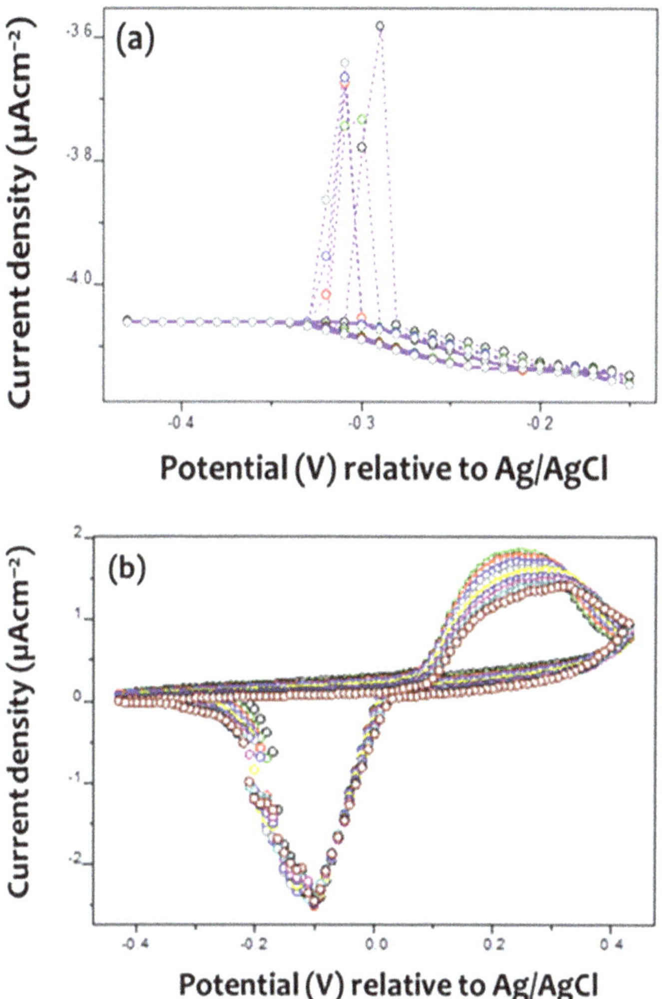

FIGURE A.3 (a) Cyclic voltammetry plots for a ZnO NR/Si (100) working electrode in 1 M H_2SO_4 at a 10 mV/s scan rate for five cycles. (b) Cyclic voltammetry plots for a ZnO NR/Si (100) working electrode in 8.5 M KOH at 10 mV/s scan rate for five cycles.

TABLE A.1

Mechanism of H_2 evolution in acidic medium

Mechanism	Metal Surface	ZnO Surface
Volmer	$H_3O^+ + e^- + M \rightarrow MH + H_2O$	$H_3O^+ + e^- \rightarrow ZnO....H + H_2O$
Heyrovsky	$H_3O^+ + e^- + MH \rightarrow M + H_2$	$H_3O^+ + e^- + ZnO...H \rightarrow ZnO + H_2O$
Tafel	$2MH \rightarrow 2M + H_2$	$2ZnO...H \rightarrow 2ZnO + H_2$

TABLE A.2
Mechanism of H_2 evolution in basic medium

Mechanism	Metal Surface	ZnO Surface
Volmer	$H_3O^+ + e^- + M \rightarrow$	$H_2O + e^- + ZnO \rightarrow$
	$MH_{ads} + OH^-$	$ZnO...H + OH^-$
Heyrovsky	$MH_{ads} + e^- + H_2O \rightarrow$	$ZnO...H + e^- + H_2O \rightarrow$
	$M + OH^- + H_2 \uparrow$	$H_2 \uparrow + ZnO + OH^-$
Tafel	$MH_{ads} \rightarrow 2H + H_2 \uparrow$	$2ZnO...H \rightarrow 2ZnO + H_2 \uparrow$

peaks is observed, $\Delta E_{shift} = 0.15$ V. This shift can be attributed to the existence of ZnO-H species after a cathodic sweep, and the evolution of H_2 occurs in an anodic sweep via recombination of ZnO-H rather than electrochemical desorption. This hypothesis is consistent with the high electronegativity of O, which would assist to hold the H atom strongly in ZnO lattice via hydrogen bonding. When an anodic potential is applied afterwards, the bonded H would start recombining with neighbouring H species, bound to ZnO, and thus promote the evolution of H_2.

A phenomenological thermodynamic analysis can be carried out for the above HER mechanisms of ZnO NR surfaces in (i) an acidic and (ii) an alkaline medium to obtain the energy required for the reaction to proceed (the activation energy, ΔG_{eq}), as well as to investigate the energies of each process occurring during the reaction.

Case (i): acidic medium

The standard exchange current density for HER on s, p, and d metal surfaces ($i_{0(HER)}$) can be written as [Harinipriya and Sangaranarayanan, 2002]:

$$i_{0(HER)} = \left(\frac{FC_H + k_b T}{Ah}\right) \exp\left\{-\frac{\Delta G_{H^+-s}}{2SN_{H^+}RT} - \frac{F\chi_e^s}{2RT} + \frac{F\xi\Phi_M}{RT}\right\}, \quad \text{Eq. (A.1)}$$

where F is Faraday constant in C/mol, C_{H^+} represents the concentration of H^+ ions in mol/litre, k_b denotes Boltzmann constant in J/K, T is the ambient temperature in K, A indicates the area of the electrode in cm^{-2}, h is the Planck's constant in Js, ΔG_{H^+-s} is the hydration energy of H^+ ions in J/mol, SN_{H^+} is the hydration number of H^+ ions, R is the Gas constant in J/K/mol, χ_e^s being the surface potential of electrons (excess energy required for the electrons to cross solution/electrode interface and vice versa) in eV,ξ is a dimensionless constant, and Φ_M is the workfunction of the electrode (in eV) under consideration. The free energy of activation (ΔG_{eq}(HER) (eV)) and heterogeneous rate constant are explained elsewhere [Harinipriya and Sangaranarayanan, 2002]. ξ is 0.17, irrespective of the cathode material. The free energy of the solution for protons (protonium ion, ΔG_{H^+-s}(eV)) and the electron surface potential in solution (χ_e^s(eV)) are reported as 9.357 eV and 0.4 eV, respectively [Trasatti, 1972, 1976;

Marcus, 1991]. The hydration number (SN_{H^+}) is given as 5, and from Figure A.3(a), peak current density for the HER on the ZnO NRs in 1 M H_2SO_4 is $-3.7\mu Acm^{-2}$.

(ii) Alkaline medium

The free energy of activation and heterogeneous rate constant are represented in details of published work [Harinipriya *et al.*, 2012]. The exchange current density can be given as:

$$i_0 = \frac{nFC_{OH} - k_b T}{Ah} \exp\left\{ -\frac{(1+\beta)E^0_{red,\frac{H^+}{H_2}}}{RT} + \frac{\Delta G^{form}_{mH}}{RT} - \frac{\Delta G^{form}_{H-H}}{RT} + \frac{pH\beta nF}{RT} \right\} \quad \text{Eq. (A.2)}$$

where $\beta = 0.5$ is the symmetry factor of the reaction, $E^0_{red,\frac{H^+}{H_2}}$ is the standard reduction potential of H^+ to H_2. From Eq. (A.1), the actual free energy change involved in the storage of H atoms in ZnO NRs can be evaluated from electrochemical data. From Figure A.3(b), if the anodic peak current is considered to be responsible for the recombination of ZnO . . . H species leading to hydrogen evolution, then $i_{0(HER)}$ is taken to be 2.5 μAcm^{-2}. Upon substituting C_{OH^-} as 8.5 and pH as 15, ΔG^{form}_{H-H} and $E^0_{red,\frac{H^+}{H_2}}$ are reported to be 4.518 eV and 0.245 V [Lide, 1987] with respect to the Ag/AgCl electrode. The free energy change involved in the bonding of hydrogen to ZnO as ZnO . . . H (ΔG^{form}_{mH}) is calculated using equation (A.2) and is approximately 192.08 6 kJ/mol (this value is approximately three times higher than the O-H . . . O type hydrogen bonding in water) [Vinogradov and Linnel, 1971].

A.2 CONCLUDING REMARKS

Vertical arrays of nanorods with preferential c-axis orientation of ZnO were grown on the native oxide on the surface of Si (100) substrates with strong NBE emission peaking at 3.176 eV. Cyclic voltammetry analysis indicated that H can be stored in ZnO NRs via the formation of ZnO–H bonding in a strongly alkaline medium.

Annexure B: MATLAB Program for Absorptance and Emittance Calculation

B.1 ABSORPTANCE CALCULATION

```
    % Caution: before running the program, kindly load data
file having first
% column as Lambda (λ) values, second as 1-R (Lambda) and
third as a reference
%*****************Input Data reading******************************%
Lambda=data(:,1);
R=data(:,2); % R represents (1-R(Lambda))
I=data(:,3);
n=size(Lambda);
Numerator=zeros(1,n(1));
for i=1:n(1)
  Numerator(i)=R(i)*I(i);     % generates complete numerator
column
  end
A=trapz(Lamda, Numerator) % Numerical integration of numerator
of absorptance expression
B=trapz(Lamda, I)     % Numerical integration of denominator
of absorptance expression
absorbance=A/B     % final absorptance value
```

B.2 EMITTANCE CALCULATION

```
    % Caution: before running the program, kindly load data
file having first
% column as Lambda values and second as R (Lambda)
%*****************Input Data reading******************************%
L=data(:,1);
Lambda=L*1e-6;
R=data(:,2);
n=size(Lambda);
T=300;
c1=3.743e-16;
c2=0.0143879;
Eb=zeros(1,n(1));
Numerator=zeros(1,n(1));
for i=1:n(1)
  Eb(i)=c1/((Lambda(i))^5*(exp(c2/(Lambda(i)*T))-1)); % generates
Eb data column
```

```
  Numerator(i)=(1-R(i))*Eb(i);     % generates complete numerator
column
  end
A=trapz(Lambda, Numerator) % Numerical integration of numera-
tor of emittance expression
B=trapz(Lambda, Eb)     % Numerical integration of denominator
of emittance expression
epsilon=A/B     % final emittance value
. . .
```

References

Abbott, A.P., Capper, G., Davies, D.L., and Rasheed, R.K., (2004), "Ionic liquid analogues formed from hydrated metal salts", *Chemistry: A European Journal*, Vol. 10, pp. 3769–3774.

Abdmouleh, Z., Alammari, R.A.M., and Gastli, A., (2015), "Review of policies encouraging renewable energy integration and best practices", *Renewable and Sustainable Energy Reviews*, Vol. 45, pp. 249–262.

Abendroth, T., Althues, H., Mäder, G., Härtel, P., Kaskel, S., and Beyer, E., (2015), "Selective absorption of Carbon Nanotube thin films for solar energy applications", *Solar Energy Materials and Solar Cells*, Vol. 143, pp. 553–556.

Agnihotri, O.P., and Gupta, B.K., (1981), *Solar Selective Surfaces*, John Wiley and Sons Inc., New York, USA.

Aidroos, D., Abdul, H., Zaidi, W., and Omar, W., (2015), "Historical development of concentrating solar power technologies to generate clean electricity efficiently – a review", *Renewable and Sustainable Energy Reviews*, Vol. 41, pp. 996–1027.

Alanyani, H., and Souto, R.M., (2003), "Research on the corrosion behavior of TiN – TiAlN multilayer coatings deposited by cathodic-arc ion plating", *Corrosion*, Vol. 59, pp. 851–854.

Ali, M.R., Nishikata, A., and Tsuru, T., (1997), "Electrodeposition of aluminum-chromium alloys from $AlCl_3$-BPC melt and its corrosion and high temperature oxidation behaviors", *Electrochimica Acta*, Vol. 42, pp. 2347–2354.

Al-Kuhaili, M.F., and Durrani, S.M.A., (2007), "Optical properties of chromium oxide thin films deposited by electron-beam evaporation", *Optical Materials*, Vol. 29, pp. 709–713.

Aman, M.M., Solangi, K.H., Hossain, M.S., Badarudin, A., Jasmon, G.B., Mokhlis, H., Bakar, A.H.A., and Kazi, S.N., (2015), "A review of safety, health and environmental (SHE) issues of solar energy system", *Renewable and Sustainable Energy Reviews*, Vol. 41, pp. 1190–1204.

Ambrosini, A., Boubault, A., Ho, C.K., Banh, L., and John, R.L., (2019), "Influence of application parameters on stability of Pyromark ® 2500 receiver coatings", *American Institute of physics Conference Proceedings*, Vol. 2126, pp. 03000.

Ambrosini, A., Lambert, T.N., Bencomo, M., Hall, A., Nathan, P., and Ho, C.K., (2011), "Improved high temperature solar absorbers for use concentrating solar power central receiver applications", Proceedings of the ASME 2011 5th International Conference on Energy Sustainabilit, pp. 1–8.

Ambrosini, A., Lambert, T.N., Boubault, A., Hunt, A., Davis, D.J., Adams, D., and Hall, A.C., (2015), "Thermal stability of oxide-based solar selective coatings for CSP central receivers", ASME 2015 9th International Conference on Sustainable ES 2015, collocated with ASME 2015 Power Conference ASME 2015 13th International Conference on Fuel Cell Science and Engineering Technology ASME 2015 Nuclear Forum 1, pp. 1–10.

Amri, A., Jiang, Z.T., Pryor, T., Yin, C.-Y., and Djordjevic, S., (2014), "Developments in the synthesis of flat plate solar selective absorber materials via sol-gel methods: A review", *Renewable and Sustainable Energy Reviews*, Vol. 36, pp. 316–328.

Anandan, C., William Grips, V.K., Rajam, K.S., Jayaram, V., and Bera, P., (2002), "Investigation of surface composition of electrodeposited black chrome coatings by X-ray photoelectron spectroscopy", *Applied Surface Science*, Vol. 191, pp. 254–260.

Andersson, A., Hunderi, O., and Granqvist, C.G., (1980), "Nickel pigmented anodic aluminum oxide for selective absorption of solar energy", *Journal of Applied Physics*, Vol. 51, pp. 754–794.

Aroutiounian, V.M., Arakelyan, V.M., and Shahnazaryan, G.E., (2005), "Metal oxide photoelectrodes for hydrogen generation using solar radiation-driven water splitting", *Solar Energy*, Vol. 78, p. 581.

Arul, R.I., (2000), "Nickel-based, binary-composite eletrocatalysts for the cathodes in the energy-efficient industrial production of hydrogen from alkaline water electrolytic cells", *Journal of Applied Electrochemistry*, Vol. 30, pp. 499–504.

Atkinson, C., Sansom, C.L., Almond, H.J., and Shaw, C.P., (2015), "Coatings for concentrating solar system – a review", *Renewable and Sustainable Energy Review*, Vol. 45, pp. 113–122.

Axelbaum, R.L., and Brandt, H., (1987), "The effect of substrate surface preparation on the optical properties of a black chrome solar absorber coating", *Solar Energy*, Vol. 39(3), pp. 233–241.

Aydoğdu, G.H., and Aydinol, M.K., (2006), "Determination of susceptibility to intergranular corrosion and electrochemical reactivation behaviour of AISI 316L type stainless steel", *Corrosion Science*, Vol. 48, pp. 3565–3583.

Barshilia, H.C., (2014), "Growth, characterization and performance evaluation Ti/AlTiAl-TiNON/AlTiO high temperature spectrally selective coatings for solar thermal power applications", *Solar Energy Materials and Solar Cells*, Vol. 130, pp. 322–330.

Barshilia, H.C., Kumar, P., Rajam, K.S., and Biswas, A., (2011), "Structure and optical properties of Ag–Al_2O_3 nanocermet solar selective coatings prepared using unbalanced magnetron sputtering", *Solar Energy Materials and Solar Cells*, Vol. 95, pp. 1707–1715.

Barshilia, H.C., Selvakumar, N., and Rajam, S., (2006a), "TiAlN/TiAlON/Si_3N_4 tandem absorber for high temperature solar selective applications", *Applied Physics Letters*, Vol. 89, pp. 19190–19191.

Barshilia, H.C., Selvakumar, N., Rajam, K.S., Sridhara Rao, D.V., Muraleedharan, K., and Biswas, A., (2006b), "TiAlN/TiAlON/Si_3N_4 tandem absorber for high temperature solar selective applications", *Applied Physics Letters*, Vol. 89, p. 191909.

Barshilia, H.C., Selvakumar, N., Rajam, K.S., and Biswas, A., (2008a), "Structure and optical properties of pulsed sputter deposited Cr_xO_y/Cr/Cr_2O_3 solar selective coatings", *Journal of Applied Physics*, Vol. 103, p. 023507.

Barshilia, H.C., Surya Prakash, M., Poojari, A., and Rajam, K., (2004), "Corrosion behavior of nanolayered TiN/NbN multilayer coatings prepared by reactive direct current magnetron sputtering process", *Thin Solid Films*, Vol. 460, pp. 133–142.

Bayati, M.R., Shariat, M.H., and Janghorban, K., (2005), "Design of chemical composition and optimum working conditions for trivalent black chromium electroplating bath used for solar thermal collectors", *Renewable Energy*, Vol. 30, pp. 2163–2178.

Bhowmik, N.C., Rahman, J., Khan, M.A.A., and Mazumder, Z.H., (2001), "Preparation of selective surfaces and determination of optimum thickness for maximum selectivity", *Renewable Energy*, Vol. 24, pp. 663–666.

Bilgin, V., Kose, S., Atay, F., Akyuz, I., (2005), "The effect of substrate temperature on the structural and some physical properties of ultrasonically sprayed CdS films", *Materials Chemistry and Physics*, Vol. 94, pp. 103–108.

Bjorksten, U., Moser, J., Gratzel, M., (1994), "Photoelectrochemical studies on nanocrystalline hematite films", *Chemistry of Materials*, Vol. 6, p. 858.

Blain, J., LeBel, C., Jacques, S., and Rheault, F., (1985), "Spectrally selective surface of Co-pigmented anodic Al_2O_3", *Journal of Applied Physics*, Vol. 58, pp. 490–494.

Blanc, P., Espinar, B., Geuder, N., Gueymard, C., Meyer, R., Pitz-Paal, R., Reinhardt, B., Renne, D., Sengupta, M., Wald, L., and Wilbert, S., (2014), "Direct normal irradiance related definitions and applications: The circumsolar issue", *Solar Energy*, Vol. 110, pp. 561–577.

Boffey, P.M., (1970), "Energy crisis: Environmental issue exacerbates power supply problem", *Science*, Vol. 168, pp. 1554–1559.

Bogaerts, W.F., and Lampert, C.M., (1983a), "Materials for photothermal solar energy conversion", *Journal of Materials Science*, Vol. 18, pp. 2847–2875.

Borzoni, J.T., (1976), "Comparison of three solar selective absorbr coatings", *Proceeduings of the American Electroplaters Society Coatings for Solar Collector Symposium*, Atlanta, Georgia, p. 89.

Boström, T.K., Wäckelgård, E., and Westin, G., (2004), "Anti-reflection coatings for solution-chemically derived nickel – Alumina solar absorbers", *Solar Energy Materials and Solar Cells*, Vol. 84, pp. 183–191.

Brett, M.J., Parsons, R.R., and Baltes, H.P., (1986), "Zinc oxide multilayers for solar sollectro coatings", *Applied Optics*, Vol. 25, pp. 2712–2714.

Brunold, S., Frei, U., Carlsson, B., Moller, K., and Kohl, M., (2000), "Accelerated life testing of solar absorber coatings: Testing procedure and results", *Solar Energy*, Vol. 68, pp. 313–323.

Burlafinger, K., Vetter, A., and Brabec, C.J., (2015), "Maximizing concentrated solar power (CSP) plant overall efficiencies by using spectral selective absorbers at optimal operation temperatures", *Solar Energy*, Vol. 120, pp. 428–438.

Butt, H.J., Cappella, B., and Kappl, M., (2005), "Force measurements with the atomic force microscope: Technique, interpretation and applications", *Surface Science Reports*, Vol. 59, pp. 1–152.

Cao, F., McEnaney, K., Chen, G., and Ren, Z., (2014a), "A review of cermet-based spectrally selective solar absorbers", *Energy & Environmental Science*, Vol. 7, pp. 1615–1627.

Cao, F., McEnaney, K., and Ren, Z., (2014b), "A review of cermet-based spectrally selective solar absorbers", *Energy and Environmental Science*, Vol. 7, pp. 1615–1627.

Cathro, K.J., (1984), "Preparation of cobalt – oxide – based by a dip – coating process", *Solar Energy Materials*, Vol. 9, pp. 433–447.

Céspedes, E., Wirz, M., Sánchez-García, J.A., Alvarez-Fraga, L., Escobar-Galindo, R., and Prieto, C., (2014), "Novel Mo – Si_3N_4 based selective coating for high temperature concentrating solar power application", *Solar Energy Materials and Solar Cells*, Vol. 122, pp. 217–225.

Chang, C.C., Huang, C.L., and Chang, C.L., (2013), "Poly(urethane)-based solar absorber coatings containing nanogold", *Solar Energy*, Vol. 91, pp. 350–357.

Chen, B., Yang, D., Charpentier, P.A., and Nikumb, S., (2008), "Optical and structural properties of pulsed deposited Ti:Al_2O_3 thin films", *Solar Energy Materials and Solar Cells*, Vol. 92, pp. 1025–1029.

Chen, C.S., Liu, C.P., Tsao, C.Y.A., and Yang, H.G., (2004), "Study of mechanical properties of PVD ZrN films, deposited under positive and negative substrate bias conditions", *Scripta Materialia*, Vol. 51, pp. 715–719.

Cheng, H.Y., Chiou, J.W., Ting, J.M., and Tzeng, Y., (2013), "Reactively co-sputter deposited a-C:H/Cr thin films: Material characteristics and optical properties", *Thin Solid Films*, Vol. 529, pp. 164–168.

Cheung, J.T., (1994), "History and fundamentals of pulsed laser deposition", in: Chrisey, D.B., and Hubler, G.K., (Eds.), *Pulsed Laser Deposition of Thin Films*, John Wiley, New York.

Chowdhury, A.K.M.S., Cameron, D.C., and Hashmi, M.S.J., (1999), "Bonding structure in carbon nitride films: Variation with nitrogen content and annealing temperature", *Surface and Coatings Technology*, Vol. 112, pp. 133–139.

Chu, S.-Y., Kline, C., Huang, M.-Q., McHenry, M.E., Cross, J., and Harris, V.G., (1999), "Preparation, characterization and magnetic properties of an ordered FeCo single crystal", *Journal of Applied Physics*, Vol. 85, p. 6031.

Conroy, M., and Amstrong, J., (2006), "A comparison of surface metrology methods", *Optical Micro- and Nanometrology in Microsystems Technology*, p. 6188.

Cox, S.F.J., Davis, E.A., Cottrell, S.P., King, P.J., Lord, J.S., Gil, J.M., *et al.*, (2001a), "Experimental confirmation of the predicted shallow donor hydrogen state in zinc oxide", *Physical Review Letters*, Vol. 86, pp. 2601–2604.

Cox, S.F.J., Davis, E.A., King, P.J.C., Gil, J.M., Alberto, H.V, Vilão, R.C., *et al.*, (2001b), "Shallow versus deep hydrogen states in ZnO and HgO", *Journal of Physics: Condensed Matter*, Vol. 13, pp. 9001–9010.

Craciun, D., Socol, G., Dorcioman, G., Stefan, N., Bourne, G., and Craciun, V., (2010), "High quality ZrC, ZrC/ZrN and ZrC/TiN thin films grown by pulsed laser deposition", *Journal of Optoelectronics and Advanced Materials*, Vol. 12, pp. 461–465.

Craighead, H.G., Howard, R.E., Sweeney, J.E., and Buhrman, R.A., (1981), "Graded-index Pt-Al_2O_3 composite solar absorbers", *Applied Physique Letters*, Vol. 39, pp. 29–31.

Crnjak Orel, Z., Leskovsek, N., Orel, B., and Hutchins, M.G., (1996), "Spectrally selective silicon paint coatings: Influence of pigment volume concentration ratio on their optical properties", *Solar Energy Materials and Solar Cells*, Vol. 40, pp. 197–204.

Cuevas, A., Martínez, L., Romero, R., Dalchiele, E.A., Marotti, R., Leinen, D., Ramos-Barrado, J.R., and Martin, F., (2014), "Electrochemically grown cobalt-alumina composite layer for solar thermal selective absorbers", *Solar Energy Materials and Solar Cells*, Vol. 130, pp. 380–386.

Cullity, B.D., (1972), *Elements of X-ray Diffraction*, Addison-Wesley, Reading, MA, p. 102.

Danilov, F.I., Protsenko, V.S., and Butyrina, T.E., (2001), "Chromium electrodeposition kinetics in solutions of Cr(III) complex ions", *Russian Journal of Electrochemistry*, Vol. 37, pp. 704–709.

Daryabegy, M., and Mahmoodpoor, A.R., (2006), "Method of manufacturing absorbing layers on copper for solar applications (I)", *Renewable Energy Organization of Iran*, Vol. 2, pp. 35–39.

Davidse, P.D., (1967a), "Rf sputter etching – a universal etch", *Vacuum*, Vol. 17, p. 165.

Davidse, P.D., (1967b), "Theory and practice of RF sputtering", *Vacuum*, Vol. 17, pp. 139–145.

Davidse, P.D., and Maisel, L.I., (1966), "Dielectric thin films through rf sputtering", *Journal of Applied Physics*, Vol. 37, p. 574.

De Giz, M.J., Tremiliosi-Filho, G., Gonzalez, E.R., Srinivasan, S., and Appleby, A.J., (1995), "The hydrogen evolution reaction on amorphous nickel and cobalt alloys", *International Journal of Hydrogen Energy*, Vol. 20, pp. 423–427.

De Hosson, J.T.M., and Cavaleiro, A., (2006), *Galileo Comes to the Surface! Nanostructured Coatings, Nanostructure Science and Technology*, Springer, New York, pp. 1–26.

de Souza Roberto, F., Padilha Janine, C., Gonc alves Reinaldo, S., de Souza Michele, O., and Rault-Berthelot, J., (2007), "Electrochemical hydrogen production from water electrolysis using ionic liquid as electrolytes: Towards the best device", *Journal of Power Sources*, Vol. 164, pp. 792–798.

Desvaux, C., Amiens, C., Fejes, P., Renaud, P., Respaud, M., Lecante, P., Snoeck, E., Chaudret, B., (2005), "Multimillimetre-large superlattices of air-stable iron-cobalt nanoparticles", *Nature Materials*, Vol. 4, pp. 750–753.

Dibari, G.A., and Turillon, P.P., (1979), "Effect of corrosion on selectivity of electrodeposited black chromium with and without nickel under layers", in: Boer, K.W., and Glenn, B.H., (Eds.), *Sun II, Proceedings of the International Solar Energy Society Congress, Atlanta*, Vol. 3, Pergamon, New York, p. 1917.

Ding, D., Cai, W., Long, M., Wu, H., and Wu, Y., (2010), "Optical, structural and thermal characteristics of Cu – $CuAl_2O_4$ hybrids deposited in anodic aluminum oxide as selective solar absorber", *Solar Energy Materials and Solar Cells*, Vol. 94, pp. 1578–1581.

Dixit, A., Sudakar, C., Thakur, J.S., Padmanabhan, K., Kumar, S., Naik, R., Naik, V.M., and Lawes, G., (2009), "Strong plasmon absorption in thin films", *Journal of Applied Physics*, Vol. 105, p. 053104.

Driver, P.M., and McCormick, P.G., (1982), "Black chrome selective surfaces I. Deposit growth and properties", *Solar Energy Materials*, Vol. 6, pp. 159–173.

Du, M., Hao, L., Mi, J., Lv, F., Liu, X., Jiang, L., and Wang, S., (2011), "Optimization design of $Ti_{0.5}Al_{0.5}N/Ti_{0.25}Al_{0.75}N/AlN$ coating used for solar selective applications", *Solar Energy Materials and Solar Cells*, Vol. 95, pp. 1193–1196.

Duffie, J.A., and Beckman, W.A., (1991), *Solar Engineering of Thermal Processes*, 2nd Edition, Wiley-Interscience, New York, USA.

El-Meligi, A.A., and Ismail, N., (2009), "Hydrogen evolution reaction of low carbon steel electrode in hydrochloric acid as a source for hydrogen production", *Iinternational Journal of Hydrogen Energy*, Vol. 34, pp. 91–97.

Endres, F., (2002), "Ionic liquids: Solvents for the electrodeposition of metals and semiconductors", *ChemPhysChem*, Vol. 3, pp. 144–154.

Endres, F., Bukowski, M., Hempelmann, R., and Natter, H., (2003), "Electrodeposition of nanocrystalline metals and alloys from ionic liquids", *Angewandte Chemie (International Ed. in English)*, Vol. 42, pp. 3428–3430.

Endres, F., MacFarlane, D.R., and Abbott, A.P., (2008), *Electrodeposition from Ionic Liquids*, Wiley-VCH.

Esposito, S., Antonaia, A., Addonizio, M.L., and Aprea, S., (2009), "Fabrication and optimisation of highly efficient cermet-based spectrally selective coatings for high operating temperature", *Thin Solid Films*, Vol. 517, pp. 6000–6006.

Est, G., and Westwood, W.D., (1998), *Handbook of Thin Film Process Technology*, IOP Publishing Ltd.

Eugénio, S., Rangel, C.M., Vilar, R., and Botelho do Rego, A.M., (2011), "Electrodeposition of black chromium spectrally selective coatings from a Cr(III)–ionic liquid solution", *Thin Solid Films*, Vol. 519, pp. 1845–1850.

Fan, J.C.C., and Spura, S.A., (1977), "Selective black absorbers using rf-sputtered Cr_2O_3/Cr Cermet films", *Applied Physics Letters*, Vol. 30, pp. 511–513.

Fan, J.C.C., and Zavracky, P.M., (1976), "Selective black absorbers using MgO/Au cermet films", *Applied Physics Letters*, Vol. 29, p. 478.

Fang, J., Tu, N., and Wei, J., (2014), "Effects of absorber emissivity on thermal performance of a solar cavity receiver", *Advances in Condensed Matter Physics*, Vol. 2014, 564639, 10 pp.

Fang, T.H., and Chang, W.J., (2004), "Nanoindentation characteristics on polycarbonate polymer film", *Microelectronics Journal*, Vol. 35, pp. 595–599.

Farooq, M., and Hutchins, M.G., (2002), "A novel design in composities of various materials for solar selective coatings", *Solar Energy Materials and Solar Cells*, Vol. 71, pp. 523–535.

Feng, J., Zhang, S., Lu, Y., Yu, H., Kang, L., Wang, X., Liu, Z., Ding, H., Tian, Y., and Ouyang, J., (2015a), "The spectral selective absorbing characteristics and thermal stability of $SS/TiAlN/TiAlSiN/Si_3N_4$ tandem absorber prepared by magnetron sputtering", *Solar Energy*, Vol. 111, pp. 350–356.

Feng, J., Zhang, S., Liu, X., Yu, H., Ding, H., Tian, Y., and Ouyang, J., (2015b), "Solar selective absorbing coatings TiN/TiSiN/SiN prepared on stainless steel substrates", *Vacuum*, Vol. 121, pp. 135–141.

Film, T., State, S., Cell, T., (1982), "Spray-pyrolysed cobalt black as a high temperature selective absorber", *Thin Solid Films*, Vol. 87, pp. 365–371.

Fontana, M.G., (2009), *Corosion Engineering*, Tata McGraw Hill Education Private Limited, New Delhi.

Galione, P.A., Baroni, A.L., Ramos-Barrado, J.R., Leinen, D., Martín, F., Marotti, R.E., and Dalchiele, E.A., (2010), "Origin of solar thermal selectivity and interference effects in nickel-alumina nanostructured films", *Surface and Coatings Technology*, Vol. 204, pp. 2197–2201.

Gaouyat, L., Mirabella, F., and Deparis, O., (2013), "Critical tuning of magnetron sputtering process parameters for optimized solar selective absorption of $NiCrO_x$ cermet coatings on aluminium substrate", *Applied Surface Science*, Vol. 271, pp. 113–117.

Gier, J.T., and Dunkle, R.V., (1958), "Selective spectral characteristics as an important factor in the efficiency of solar collectors", in: *Transactions of the Conference on the Use of Solar Energy*, Vol. 2, University of Arizona Press, Tucson, p. 41.

Glaude, A.S., Bousquet, I., Thomas, L., and Flamant, G., (2013), "Optical modeling of multi-layered coatings based on SiC(N)H materials for their potential use as high-temperature solar selective absorbers", *Solar Energy Materials and Solar Cells*, Vol. 117, pp. 315–323.

Gong, D., Liu, H., Luo, G., Zhang, P., Cheng, X., Yang, B., Wang, Y., Min, J., Wang, W., Chen, S., Cui, Z., Li, K., and Hu, L., (2015), "Thermal aging test of AlCrNO-based solar selective absorbing coatings prepared by cathodic arc plating", *Solar Energy Materials and Solar Cells*, Vol. 136, pp. 167–171.

Granqvist, C.G., and Niklasson, G.A., (1977), "Selective absorption of solar energy in ultrafine chromium particles", *Applied Physics Letters*, Vol. 31, pp. 665–666.

Gray, M.H., Tirawat, R., Kessinger, K., and Ndione, P.F., (2015), "High temperature performance of high-efficiency, multi-layer solar selective coatings for tower applications", *Energy Procedia*, Vol. 69, pp. 398–404.

Griffiths, D.J., (1999), *Introduction to Electrodynamics*, 3rd Edition, Prentice Hall, New Jersey, USA.

Grimmer, D.P., and Collier, R.K., (1981), "Black-chrome solar-selective coatings electrodeposited on metallized glass tubes", *Solar Energy*, Vol. 26, pp. 467–469.

Grips, V.K.W., Barshilia, H.C., Selvi, V.E., and Rajam, K.S., (2006), "Electrochemical behavior of single layer CrN, TiN TiAlN coatings and nanolayered TiAlN/CrN multilayer coatings prepared by reactive direct current magnetron sputtering", *Thin Solid Films*, Vol. 514, pp. 204–211.

Gupta, V., Patra, M.K., Shukla, A., Saini, L., Songara, S., Jani, R., Vadera, S.R., and Kumar, N., (2014), "Synthesis and investigations on microwave absorption properties of core – shell FeCo(C) alloy nanoparticles", *Science of Advanced Materials*, Vol. 6, pp. 1196–1202.

Hall, A., (2012), "Solar selective coatings for concentrating solar power central receivers", *Advances in Materials and Processing*, Vol. 170, pp. 28–32.

Hamid, Z.A., (2009), "Electrodeposition of black chromium from environmentally electrolyte based on trivalent chromium salt", *Surface and Coatings Technology*, Vol. 203, pp. 3442–3449.

Han, J., Wan, F., Zhu, Z., Liao, Y., Ji, T., Ge, M., and Zhang, Z., (2005), "Shift in low-frequency vibrational spectra of transition-metal zirconium compounds", *Applied Physics Letters*, Vol. 87, p. 172107.

Hao, L., Wang, S., Jiang, L., Liu, X., Li, H., and Li, Z., (2009), "Preparation and thermal stability on non-vacuum high temperature solar selective absorbing coatings", *Chinese Science Bulletin*, Vol. 54, pp. 1451–1454.

Harding, G.L., (1979), "Alternative grading profile for sputtered solar selective surface", *Journal of Vacuum Science and Technology*, Vol. 16, pp. 2111–2113.

Harinipriya, S., and Sangaranarayanan, M.V., (2002), "Influence of the work function on electron transfer processes at metals: Application to the hydrogen evolution reaction", *Langmuir*, Vol. 18, p. 5572.

Harinipriya, S., Usmani, B., Rogers, D.J., Sandana, V.E., Hosseini Teherani, F., Lusson, A., Bove, P., Drouhin, H.-J., and Razeghi, M., (2012), "ZnO nanorod electrodes for hydrogen evolution & storage", *Proceedings of SPIE*, Vol. 8263.

Harrington, D.A., and Conway, B.E., (1987), "AC impedance of faradaic reactions involving electrosorbed intermediates. Kinetic theory", *Electrochimica Acta*, Vol. 32, p. 1703.

Ho, C.K., (2017), "Advances in central receivers for concentrating solar applications", *Solar Energy*, Vol. 152, pp. 38–56.

Ho, C.K., Mahoney, A.R., and Lambert, T.N., (2016), "Characterization of Pyromark 2500 paint for high-temperature solar receivers", *Journal of Solar Energy Engineering*, Vol. 136, pp. 2014–2017.

Hofmann, D.M., Hofstaetter, A., Leiter, F., Zhou, H., Henecker, F., Meyer, B.K., *et al.*, (2002), "Hydrogen: A relevant shallow donor in zinc oxide", *Physical Review Letters*, Vol. 88, p. 045504.

Hogg, S.W., and Smith, G.B., (1977), "The unusual and useful optical properties of electrodeposited chrome-black films", *Journal of Physics D: Applied Physics*, Vol. 10, pp. 1863–1869.

Holland, L., (1956), *Vacuum Deposition of Thin Films*, Chapman & Hall, London, UK.

Holloway, P.H., Shanker, K., Pettit, R.B., and Sowell, R.R., (1980), "Oxidation of electrodeposited black chrome selective solar absorber films", *Thin Solid Films*, Vol. 72, pp. 121–128.

Hu, W.K., (2000), "Electrocatalytic properties of new electrocatalysts for hydrogen evolution in alkaline water electrolysis", *International Journal of Hydrogen Energy*, Vol. 25, pp. 111–118.

Huang, J.-H., Ouyang, F.-Y., and Yu, G.-P., (2007a), "Effect of film thickness and Ti interlayer on the structure and properties of nanocrystalline TiN thin films on AISI D2 steel", *Surface and Coatings Technology*, Vol. 201, pp. 7043–7053.

Huang, J.-H., Chang, K.-H., and Yu, G.-P., (2007b), "Synthesis and characterization of nanocrystalline ZrN_xO_y thin films on Si by ion plating", *Surface and Coatings Technology*, Vol. 201, pp. 6404–6413.

Huang, J.H., Lin, T.C., and Yu, G.P., (2009), "Heat treatment induced phase separation and phase transformation of ZrN_xO_y thin films deposited by ion plating", *Surface and Coatings Technology*, Vol. 203, pp. 3491–3500.

Huang, J.H., Lin, T.C., and Yu, G.P., (2011a), "Phase transition and mechanical properties of ZrN_xO_y thin films on AISI 304 stainless Steel", *Surface and Coatings Technology*, Vol. 206, pp. 107–116.

Huang, J.H., Lin, T.C., and Yu, G.P., (2011b), "Structure evolution and mechanical properties of ZrN_xO_y thin film deposited on Si by magnetron sputtering", *Surface and Coatings Technology*, Vol. 205, pp. 5093–5102.

Huang, J.-H., Tsai, Z.-E., and Yu, G.-P., (2008), "Mechanical properties and corrosion resistance of nanocrystalline ZrN_xO_y coatings on AISI 304 stainless steel by ion plating", *Surface and Coatings Technology*, Vol. 202, pp. 4992–5000.

Ignatiev, A., O'Neill, P., Doland, C., and Zajac, G., (1979a), "Microstructure dependence of the optical properties of solar-absorbing black chrome", *Applied Physics Letters*, Vol. 34, p. 42.

Ignatiev, A., O'Neill, P., and Zajac, G., (1979b), "The surface microstructure optical properties relationship in solar absorbers: Black chrome", *Solar Energy Materials*, Vol. 1, pp. 69–79.

Inal, O.T., Valayapetre, M., Murr, L.E., and Torma, A.E., (1981), "Microstructural and mechanical property evaluation of black-chrome coated solar collectors – II", *Solar Energy Materials*, Vol. 4(3), pp. 333–358.

Iqbal, M., (1983), *An Introduction to Solar Radiation*, Academic Press, Ontario, Canada.

Ithurbide, A., Frateur, I., Galtayries, A., and Marcus, P., (2007), "XPS and flow-cell EQCM study of albumin adsorption on passivated chromium surfaces: Influence of potential and pH", *Electrochimica Acta*, Vol. 53, pp. 1336–1345.

Jafari, S, and Rozati, S.M., (2011), "Characterization of black chrome film prepared by electroplating technique", World Renewable Energy Congress-2011, Sweden.

Japelj, B., Vuk, A.Š., Orel, B., Perše, L.S., Jerman, I., and Kovač, J., (2008), "Preparation of a TiMEMO nanocomposite by the sol – gel method and its application in coloured thickness insensitive spectrally selective (TISS) coatings", *Solar Energy Materials and Solar Cells*, Vol. 92, pp. 1149–1161.

Jaworske, D.A., (2002), *Duraibility of Solar Selective Coatings Simulated Space Enviroment*, NASA Glenn Research Center, Cleveland, OH.

Joanna, P., and Antoni, B., (2007), "Production and electrochemical characterization of Ni-based composite coatings containing titanium, vanadium or molybdenum powders", *Surface and Coatings Technology*, Vol. 201, pp. 6478–6483.

Joint Committee on Powder Diffraction Standards, File No. 00-065-6829.

Joint Committee on Powder Diffraction Standards, File No. 01-074-7045.

Joint Committee on Powder Diffraction Standards, File No. 01-077-7591.

Joly, M., Antonetti, Y., Python, M., Gonzalez, M., Gascou, T., Scartezzini, J.-L., and Schüler, A., (2013), "Novel black selective coating for tubular solar absorbers based on a sol – gel method", *Solar Energy*, Vol. 94, pp. 233–239.

Kalogirou, S.A., (2004), "Solar thermal collectors and applications", *Progress in Energy and Combustion Science*, Vol. 30, pp. 231–295.

Kaluza, L., Orel, B., Drazic, G., and Kohl, M., (2001), "Sol-gel derived $CuCoMnO_x$spinel coatings for solar absorbers: Structural and optical properties", *Solar Energy Materials and Solar Cells*, Vol. 70, pp. 187–201.

Kanu, S.S., and Binions, R., (2010), "Thin film for solar control applications", *Proceedings of the Royal Society A: Mathematical, Physical and Engineering Sciences*, Vol. 466, pp. 19–44.

Karas, D.E., Byun, J., Moon, J., and Jose, C., (2018), "Solar energy materials and solar cells copper-oxide spinel absorber coatings for high-temperature concentrated solar power systems", *Solar Energy Materials and Solar Cells*, Vol. 182, pp. 321–330.

Karthick Kumar, S., Suresh, S., Murugesan, S., and Raj, S.P., (2013), "CuO thin films made of nanofibers for solar selective absorber applications", *Solar Energy*, Vo. 94, pp. 299–304.

Katumba, G., Makiwa, G., Baisitse, T.R., Olumekor, L., Forbes, A., and Wäckelgård, E., (2008a), "Solar selective absorber functionality of carbon nanoparticles embedded in SiO_2, ZnO and NiO matrices", *Physica Status Solidi (C)*, Vol. 5(2), pp. 549–551.

Katumba, G., Olumekor, L., Forbes, A., Makiwa, G., Mwakikunga, B., Lu, J., and Wäckelgård, E., (2008b), "Optical, thermal and structural characteristics of carbon nanoparticles embedded in ZnO and NiO as selective solar absorbers", *Solar Energy Materials and Solar Cells*, Vol. 92, pp. 1285–1292.

Katzen, D., Levy, E., and Mastai, Y., (2005), "Thin films of silica – carbon nanocomposites for selective solar absorbers", *Applied Surface Science*, Vol. 248, pp. 514–517.

Kelly, P.J., and Arnell, R.D., (2000), "Magnetron sputtering: A review of recent developments and applications", *Vacuum*, Vol. 56, pp. 159–172.

Kennedy, C.E., (2002), "Review of mid-to high-temperature solar absorber materials", NREL/TP-520-31267, National Renewable Energy Laboratory, Colorado.

Kennedy, C.E., (2014), "High temperature solar selective coatings", US Patent No. 8893711, 25.

Keramidas, V.G., and White, W.B., (1974), "Raman scattering study of the crystallization and phase transformations of ZrO_2", *Journal of the American Ceramic Society*, Vol. 57, pp. 22–24.

Kılıç, C., and Zunger, A., (2002), "N-type doping of oxides by hydrogen", *Applied Physics Letters*, Vol. 81, p. 73.

Kissinger, P.T., and Heineman, W.R., (1996a), *Laboratory Techniques in Electroanalytical Chemistry*, 2nd Edition, Marcel Dekker, Inc., New York.

Kissinger, P.T., and Heineman, W.R., (1996b), *Laboratory Technique in Electrochemistry*, 2nd Edition, Marcel Dekker, Inc., Basel, NY.

Kittel, C., (1995), *Introduction to Solid State Physics*, 7th Edition, John Wiley & Sons, Inc., New York, Chichester, Brisbane, Toronto, Singapore.

Klochko, N.P., Klepikova, K.S., Tyukhov, I.I., Myagchenko, Y.O., Melnychuk, E.E., Kopach, V.R., Khrypunov, G.S., Lyubov, V.M., Kopach, A.V., Starikov, V.V., and Kirichenko, M.V., (2015), "Zinc oxide-nickel cermet selective coatings obtained by sequential electrodeposition", *Solar Energy*, Vol. 117, pp. 1–9.

Köhl, M., Heck, M., Brunold, S., Frei, U., Carlsson, B., and Möller, K., (2004), "Advanced procedure for the assessment of the lifetime of solar absorber coatings", *Solar Energy Materials and Solar Cells*, Vol. 84, pp. 275–289.

Konttinen, P., Lund, P.D., and Kilpi, R.J., (2003), "Mechanically manufactured selective solar absorber surfaces", *Solar Energy Materials and Solar Cells*, Vol. 79, pp. 273–283.

Kubaschewski, O., (1993), *Materials Thermochemistry*, Pergamon Press, Oxford.

Kubelka, P., and Munk, F., (1931), "Ein Beitrag zur Optik der. Farbanstriche", *Z Technical Physics (Leipzig)*, Vol. 12, pp. 593–601.

Kunič, R., Mihelčič, M., Orel, B., Slemenik Perše, L., Bizjak, B., Kovač, J., and Brunold, S., (2011), "Life expectancy prediction and application properties of novel polyurethane based thickness sensitive and thickness insensitive spectrally selective paint coatings for solar absorbers", *Solar Energy Materials and Solar Cells*, Vol. 95, pp. 2965–2975.

Kurt, R., Sanjines, R., Karimi, A., and Lévy, F., (2000), "Structural and mechanical properties of CN_x thin films prepared by magnetron sputtering", *Diamond and Related Materials*, Vol. 9, pp. 566–572.

Lampert, C.M., (1978), "Microstructure of a black chrome solar selective absorber", SPIE 22nd Institutional Technical Symposium, San Diego, CA, August 28–31.

Lampert, C.M., (1979a), "Coatings for enhanced photothermal energy collection", *Solar EnergyeMaterials*, Vol. 1, pp. 319–341.

Lampert, C.M., (1979b), "Microstructure and optical properties of black chrome before and after exposure to high temperatures", Second Annual 1 Conference on absorber Surfaces for Solar Recievers, Boulder, Colorado, January 24–25.

Lampert, C.M., (1980), "Metallurgical analysis and high temperature degration of the black chrome solar selective absorber", *Thin Soid Film.*

Lampert, C.M., and Washburn, J., (1979), "Microstructure of a black chrome solar selective absorber", *Solar Energy Materials*, Vol. 1, pp. 81–92.

Lanxner, M., and Elgat, Z., (1990), "Solar select coating for high service temperature produced by plasma sputtering", *Proceeding of the Society of Photo-Optical Instrumentation Engineers*, Vol. 1272, pp. 240–249.

Larrouturou, F., Caliot, C., and Flamant, G., (2016), "ScienceDirect influence of receiver surface spectral selectivity on the solar-to-electric efficiency of a solar tower power plant", *Solar Energy*, Vol. 130, pp. 60–73.

Lazarov, M.P., and Mayer, I.V., (1997), US Patent No. 5670248.

Lazarov, M.P., and Mayer, I.V., (1998), US Patent No. 5776556.

Leach, R.K., (2011), *Measurement Good Practice Guide: The Measurement of Surface Texture Using Stylus Instruments*, National Physical Laboratory, United Kingdom.

Lee, C.W., Eom, S.W., Sathiyanarayanan, K., and Yun, M.S., (2006), "Preliminary comparative studies of zinc and zinc oxide electrodes on corrosion reaction and reversible reaction for zinc/air fuel cells", *Electrochimica Acta*, Vol. 52, pp. 1588–1591.

Lee, K.D., (2006), "Characterization of Cr-O cermet solar selective coatings deposited by using direct-current magnetron sputtering technology", *Journal of the Korean Physical Society*, Vol. 49, pp. 187–194.

Lee, K.D., Jung, W.C., and Kim, J.H., (2000), "Thermal degradation of black chrome coatings", *Solar Energy Materials and Solar Cells*, Vol. 63, pp. 125–137.

Lee, S.J., Cho, J.-H., Lee, C., Cho, J., Kim, Y.-R., and Park, J.K., (2011), "Synthesis of highly magnetic graphite-encapsulated FeCo nanoparticles using a hydrothermal process", *Nanotechnology*, Vol. 22, p. 375603.

Leng, Y., (2013), *Materials Characterization: Introduction to Microscopic and Spectroscopic Methods*, 2nd Edtion, Wiley-CVH Verlag GmbH & Co., KGaA.

Li, L., (2000), "AC anodization of aluminum, electrodeposition of nickel and optical property examination", *Solar Energy Materials and Solar Cells*, Vol. 64, pp. 279–289.

Lide, D.R., (1987), *CRC Handbook of Chemistry and Physics*, 68th Edition, CRC Press Inc., Florida, USA.

Lind, M.A., Pettit, R.B., and Masterson, K.D., (1980), "The sensitivity of solar transmittance reflectance and absorptance to selected averaging procedure and solar irradiance distributions", *Transactions on ASME*, Vol. 102, pp. 34–40.

Lior, N., (2008), "Energy resources and use: The present situation and possible paths to the future", *Energy*, Vol. 33, pp. 842–857.

Lira-Cantú, M., Sabio, A.M., Brustenga, A., and Gómez-Romero, P., (2005), "Electrochemical deposition of black nickel solar absorber coatings on stainless steel AISI316L for thermal solar cells", *Solar Energy Materials and Solar Cells*, Vol. 87, pp. 685–694.

Liu, C., Leyland, A., Bi, Q., and Matthews, A., (2001), "Corrosion resistance of multi-layered plasma-assisted physical vapour deposition TiN and CrN coatings", *Surface and Coatings Technology*, Vol. 141, pp. 164–173.

Liu, H.D., Wan, Q., Lin, B.Z., Wang, L.L., Yang, X.F., Wang, R.Y., Gong, D.Q., Wang, Y.B., Ren, F., Chen, Y.M., Cheng, X.D., and Yang, B., (2014a), "The spectral properties and thermal stability of CrAlO-based solar selective absorbing nanocomposite coating", *Solar Energy Materials and Solar Cells*, Vol. 122, pp. 226–232.

Liu, Y., Wang, Z., Lei, D., and Wang, C., (2014b), "A new solar spectral selective absorbing coating of SS–(Fe_3O_4)/Mo/TiZrN/TiZrON/SiON for high temperature application", *Solar Energy Materials and Solar Cells*, Vol. 127, pp. 143–146.

Liu, H.D., Wan, Q., Xu, Y.R., Luo, C., Chen, Y.M., Fu, D.J., Ren, F., Luo, G., Cheng, X.D., Hu, X.J., and Yang, B., (2015), "Long-term thermal stability of CrAlO-based solar selective absorbing coating in elevated temperature air", *Solar Energy Materials and Solar Cells*, Vol. 134, pp. 261–267.

Liu, S.-Y., Perng, Y.-H., and Ho, Y.-F., (2013), "The effect of renewable energy application on Taiwan buildings: What are the challenges and strategies for solar energy exploitation?", *Renewable and Sustainable Energy Reviews*, Vol. 28, pp. 92–106.

Liu, Y., Wang, C., and Xue, Y., (2012), "The spectral properties and thermal stability of NbTiON solar selective absorbing coating", *Solar Energy Materials and Solar Cells*, Vol. 96, pp. 131–136.

López-Herraiz, M., Bello, A., Martinez, N., and Gallas, M., (2017), "Effect of the optical properties of the coating of a concentrated solar power central receiver on its thermal ef fi ciency", *Solar Energy Materials and Solar Cells*, Vol. 159, pp. 66–72.

Lou, H.H., and Huang, Y., (2006), "Electroplating", *Encyclopedia of Chemical Processing*. https://doi.org/10.1081/E-ECHIP-120007747.

Mabon, J.C., Inal, O.T., and Singh, A.J., (1982), "An investigation of deposition parameter dependence of optical properties, microstructure and thermal stability of black chrome selective surfaces", *Solar Energy Materials*, Vol. 7, pp. 359–376.

Marcus, Y, (1991), "Thermodynamics of solvation of ions. Part 5. – Gibbs free energy of hydration at 298.15 K", *Journal of the Chemical Society, Faraday Transactions*, Vol. 87, pp. 2995–2999.

Masters, G.M., (2004), *Renewable and Efficient Electric Power System*, John Wiley & Sons, Inc., Hoboken, NJ, USA.

Mattox, D.M., (1997), *Handbook of Physical Vapor Deposition (PVD) Processing*, Noyes Publications, Westwood, NJ, USA.

McClanahan, E.D., and Laegried, N., (1991), *Sputtering by Particle Bombardment III*, Springer Verlag, Berlin, p. 339.

McDonald, G., (1980), "A preliminary study of solar selective coating system using a black cobalt oxide for high temperature solar collectors", *Thin Solid Films*, Vol. 72, pp. 83–87.

McDonald, G.E., (1974), *Refinemnet in Black Chrome for Use as a Solar Selective Coating*, Lewies Research Center, National Aeronautics and Space Administration, Washington, DC.

McDonald, G.E., (1975), "Spectral reflectance properties of black chrome for use as a solar selective coating", *Solar Energy*, Vol. 17, pp. 119–122.

McDonald, G.E., (1980), "A preliminary study of solar selective coating system using a black cobalt oxide for high temperature solar collectors", *Thin Solid Films*, Vol. 72, pp. 83–87.

McDonald, G.E., and Curtis, H.B., (1975), *Variation of Solar-Selective Properties of Black Chrome with Plating Time*, Lewies Research Center, NASA-TM-X-717331.

McDonald, G.E., and Curtis, H.B., (1976), Technical Report No. NASA-TMX-73498.

Mckenzie, D.R., (1979), "Gold, silver, chromium and copper cermet selective surfaces for evacuated solar collectors", *Applied Physics Letters*, Vol. 34, pp. 25–28.

Meinel, A.B., and Meinel, M.P., (1977), *Applied Solar Energy*, Addison-Wesley.

Mekhilef, S., Saidur, R., and Safari, A., (2011), "A review on solar energy use in industries", *Renewable and Sustainable Energy Reviews*, Vol. 15, pp. 1777–1790.

Merchán, R.P., Santos, M.J., Medina, A., and Hernández, A.C., (2022), "High temperature central tower plants for concentrated solar power: 2021 overview", *Renewable and Sustainable Energy Reviews*, Vol. 155, p. 111828.

Metikos-Hukovic, M., Grubac, Z., Radic, N., and Tonejc, A., (2006), "Sputter deposited nanocrystalline Ni and Ni – W films as catalysts for hydrogen evolution", *Journal of Molecular Catalysis A: Chemical*, Vol. 249, pp. 172–180.

Möller, T., and Hönicke, D., (1998), "Solar selective properties of electrodeposited thin layers on aluminium", *Solar Energy Materials and Solar Cells*, Vol. 54, pp. 397–403.

Moon, J., Kim, T.K., VanSaders, B., Choi, C., Liu, Z., Jin, S., and Chen, R., (2015), "Black oxide nanoparticles as durable solar absorbing material for high-temperature concentrating solar power system", *Solar Energy Materials and Solar Cells*, Vol. 134, pp. 417–424.

Moon, J., Lu, D., VanSaders, B., Kim, T.K., Kong, S.D., Jin, S., Chen, R., and Liu, Z., (2014), "High performance multi-scaled nanostructured spectrally selective coating for concentrating solar power", *Nano Energy*, Vol. 8, pp. 238–246.

Mukhopadhyay, I., Aravinda, C.L., Borissov, D., and Freyland, W., (2005), "Electrodeposition of Ti from $TiCl_4$ in the ionic liquid l-methyl-3-butyl-imidazolium bis (trifluoro methyl sulfone) imide at room temperature: Study on phase formation by in situ electrochemical scanning tunneling microscopy", *Electrochimica Acta*, Vol. 50, pp. 1275–1281.

Murr, L.E., Inal, O.T., and Valayapetre, M., (1980), "Characterization of selective solar absorber microstructures: Electron microscope studies", *Thin Solid Films*, Vol. 72, pp. 111–120.

Musil, J., Baroch, P., Vlček, J., Nam, K.H., and Han, J.G., (2005), "Reactive magnetron sputtering of thin films: Present status and trends", *Thin Solid Films*, Vol. 475, pp. 208–218.

Nelson, V.C., (2020), "Concentrating solar power", *Introduction to Renewable Energy*, pp. 149–164.

Niklasson, G.A., and Granqvist, C.G., (1982), "Dielectric function of coevaporated Co-Al_2O_3 cermet films", *Applied Physics Letters*, Vol. 41, p. 773.

Niklasson, G.A., and Granqvist, C.G., (1983), "Solar absorptance and thermal emittance of co-evaporated Co-Al_2O_3 cermet films", *Solar Energy Materials*, Vol. 7, pp. 501–510.

Niklasson, G.A., and Granqvist, C.G., (1984), "Optical properties and solar selectivity of co-evaporated Co-Al_2O_3 composite films", *Journal of Applied Physics*, Vol. 55, p. 3382.

Nou, J., Chauvin, R., Thil, S., and Grieu, S., (2016), "A new approach to the real-time assessment of the clear-sky direct normal irradiance", *Applied Mathematical Modelling*, Vol. 40, pp. 7245–7264.

Nuru, Z.Y., Arendse, C.J., Muller, T.F.G., Khamlich, S., and Maaza, M., (2014), "Thermal stability of electron beam evaporated Al_xO_y/Pt/Al_xO_y multilayer solar absorber coatings", *Solar Energy Materials and Solar Cells*, Vol. 120, pp. 473–480.

Nuru, Z.Y., Arendse, C.J., Muller, T.F.G., and Maaza, M., (2012a), "Structural and optical properties of Al_xO_y/Pt/Al_xO_y multilayer absorber", *Materials Science and Engineering: B*, Vol. 177, pp. 1194–1199.

Nuru, Z.Y., Arendse, C.J., Nemutudi, R., Nemraoui, O., and Maaza, M., (2012b), "Pt – Al_2O_3 nanocoatings for high temperature concentrated solar thermal power applications", *Physica B: Condensed Matter*, Vol. 407, pp. 1634–1637.

Nuru, Z.Y., Msimanga, M., Muller, T.F.G., Arendse, C.J., Mtshali, C., and Maaza, M., (2015), "Microstructural, optical properties and thermal stability of MgO/Zr/MgO multilayered selective solar absorber coatings", *Solar Energy*, Vol. 111, pp. 357–363.

Oliver, W.C., and Pharr, G.M., (1992), "An improved technique for determining hardness and elastic modulus using load and displacement sensing indentation experiments", *Journal of Materials Research*, Vol. 7, pp. 1564–1583.

Oliver, W.C., and Pharr, G.M., (2004), "Measurement of hardness and elastic modulus by instrumented indentation Advances in understanding and refinements to methodology", *Materials Research Society*, Vol. 19, pp. 3–20.

Orel, B., Crnjak Orel, Z., Jerman, R., and Radoczy, I., (1990), "Coil-coating paints for solar collector panels – II. FT-IR spectroscopic investigations", *Solar & Wind Technology*, Vol. 7, pp. 713–717.

Orel, B., Crnjak Orel, Z., Krainer, A., and Hutchins, M.G., (1991), "FTIR spectroscopic investigations and t the thermal stability of thickness sensitive spectrally selective (TSSS) paint coatings", *Solar Energy Materials*, Vol. 22, pp. 259–279.

Paunovic, M., and Schlesinger, M., (2006), *Fundamental of Electrochemical Deposition*, 2nd Edition, John Wiley & Sons, Inc., Hoboken, NJ, USA.

Perissi, I., Bardi, U., Caporali, S., and Lavacchi, A., (2006), "High temperature corrosion properties of ionic liquids", *Corrosion Science*, Vol. 48, pp. 2349–2362.

Petch, N.J., (1953), "The cleavage strength of polycrystals", *Tetsu To Hagane*, Vol. 174, pp. 25–28.

Peterson, R.E., and Ramsey, J.W., (1975), "Thin film coatings in solar–thermal power systems", *Journal of Vacuum Science and Technology*, Vol. 12, p. 174.

Pettit, R.B., and Sowell, R.R., (1976), "Solar absorptance and emittance properties of several solar coatings", *Journal of Vacuum Science and Technology*, Vol. 13, p. 596.

Poddar, P., Wilson, J.L., Srikanth, H., Ravi, B.G., Wachsmuth, J., and Sudarshan, T.S., (2004), "Grain size influence on soft ferromagnetic properties in Fe-Co nanoparticles", *Materials Science and Engineering B: Solid-State Materials for Advanced Technology*, Vol. 106, pp. 95–100.

Pradhan, A.A., and Shah, S.I., (2002), "High deposition rate ractive sputtering with hollow cathode", 45th Annual Technical Conference Proceeding of the Society of Vacuum Coaters, pp. 96–100.

Prasad, A.A., Taylor, R.A., and Kay, M., (2015), "Assessment of direct normal irradiance and cloud connections using satellite data over Australia", *Applied Energy*, Vol. 143, pp. 301–311.

Prasad, G.S.S, Mohan, S., and Vasu, K.I., (1980), "Microstructural characterization of an electrodeposited black chromium surface", *Journal of the Electrochemical Society India*, Vol. 29, p. 82.

Qing, H., Huiren, L., Jianshe, C., Xin, L., and Xujun, W., (2004), "Study of amorphous Ni – S – Co alloy used as hydrogen evolution reaction cathode in alkaline medium", *International Journal of Hydrogen Energy*, Vol. 29, pp. 243–248.

Qing, H., Huiren, L., Jianshe, C., and Xujun, W., (2003), "A study of the electrodeposited Ni – S alloys as hydrogen evolution reaction cathodes", *International Journal of Hydrogen Energy*, Vol. 28, pp. 1207–1212.

Quintana, J., and Sebastian, P.J., (1994), "The influence of various substrate treatments on morphology and selective absorber characteristics of electrochemical black chrome", *Solar Energy Materials and Solar Cells*, Vol. 33, pp. 465–474.

Quinten, M., (2011), *Optical Properties of Nanoparticle Systems*, Mie and Beyo, Wiley-VCH Publisher.

Rajagopalan, I., Gripps, W., Vasudevan, N., Rajagopalan, S.R., and Ramaseshan, S., (1978), "A new process for black coatings useful in harnessing solar energy I – a room temperature black chromium plating bath", *Proceedings of the International Solar Energy Society Conference, New Delhi, India, January 16–21*, Vol. 2, p. 855.

Rajan, T.V., Sharma, C.P., and Sharma, A., (1992), *Heat Treatment: Principles and Technique*, Prentence Hall.

Rau Greg, H., (2004), "Possible use of Fe/CO_2 fuel cells for CO_2 mitigation plus H_2 and electricity production", *Energy Conversion and Management*, Vol. 45, pp. 2143–2152.

Rebouta, L., Pitães, A., Andritschky, M., Capela, P., Cerqueira, M.F., Matilainen, A., and Pischow, K., (2012), "Optical characterization of TiAlN/TiAlON/SiO_2 absorber for solar selective applications", *Surface and Coatings Technology*, Vol. 211, pp. 41–44.

Rebouta, L., Sousa, A., Capela, P., Andritschky, M., Santilli, P., Matilainen, A., Pischow, K., Barradas, N.P., and Alves, E., (2015), "Solar selective absorbers based on Al_2O_3:W cermets and AlSiN/AlSiON layers", *Solar Energy Materials and Solar Cells*, Vol. 137, pp. 93–100.

Reiss, H., (1981), "Polymer layers protective against corrosion and abrasion, transparent in the infrared and with low thermal emissivity on electrodeposited selectively absorbing surfaces in solar collectors", *Thin Solid Films*, Vol. 75, pp. 261–269.

Richtmyer, F.K., and Kennard, E.H., (1947), *Introduction to Modern Physics*, 4th Edition, McGraw-Hill, New York, USA.

Riedl, H., Koller, C.M., Munnik, F., Hutter, H., Martin, F.M., Rachbauer, R., *et al.*, (2016), "Influence of oxygen impurities on growth morphology, structure and mechanical properties of Ti-Al-N thin films", *Thin Solid Films*, Vol. 603, pp. 39–49.

Ritchie, I.T., Sharma, S.K., Valignat, J., and Spitz, J., (1979), "Thermal degradation of chromium black solar selective absorbers", *Solar Energy Materials*, Vol. 2, pp. 167–176.

Roberge, P.R., (2008), *Corrosion Engineering: Principles and Practice*, McGraw-Hill.

Rogers, D.J., Sandana, V.E., Hosseini Teherani, F., McClintock, R., Razeghi, M., and Drouhin, H.J., (2011), "Amorphous ZnO films grown by room temperature pulsed laser deposition on paper and mylar for transparent electronics applications", *Proceedings of SPIE*, Vol. 7940, p. 79401K-1.

Roro, K.T., Tile, N., and Forbes, A., (2012a), "Preparation and characterization of carbon/nickel oxide nanocomposite coatings for solar absorber applications", *Applied Surface Science*, Vol. 258, pp. 7174–7180.

Roro, K.T., Tile, N., Mwakikunga, B., Yalisi, B., and Forbes, A., (2012b), "Solar absorption and thermal emission properties of multiwall carbon nanotube/nickel oxide nanocomposite thin films synthesized by sol – gel process", *Materials Science and Engineering: B*, Vol. 177, pp. 581–587.

Rosalbino, F., Maccio, D., Angelini, E., Saccone, A., and Delfino, S., (2005), "Electrocatalytic properties of Fe – R (R = rare earth metal) crystalline alloys as hydrogen electrodes in alkaline water electrolysis", *Journal of Alloys and Compounds*, Vol. 403, pp. 275–282.

Rossmeisl, J., Qu, Z.W., Zhu, H., Kroes, G.J., and Nørskov, J.K., (2007), "Electrolysis of water on oxide surfaces", *Journal of Electroanalytical Chemistry*, Vol. 607, pp. 83–89.

Rossnagel, S.M., (1991), *Thin Film Processes II*, Vossen, J.L., and Kern, W., (Eds.), Academic Press, New York.

RREDC, (2015), http://rredc.nrel.gov/solar/spectra/am1.5/.

Rubin, E.B., Chen, Y., and Chen, R., (2019), "Optical properties and thermal stability of Cu spinel oxide nanoparticle solar absorber coatings", *Solar Energy Materials and Solar Cells*, Vol. 195, pp. 81–88.

Salmi, J., Bonino, J.P., and Bes, R.S., (2000), "Nickel pigmented anodized aluminum as solar selective absorbers", *Journal of Materials Science*, Vol. 35, pp. 1347–1351.

Sandana, V.E., Rogers, D.J., Hosseini Teherani, F., Orsal, G., Molinari, M., Troyon, M., Gautier, S., Moudakir, T., Abid, M., Ougazzaden, A., Largeteau, A., Demazeau, G., Drouhin, H.-J., Scott, C., and Razeghi, M., (2013), "Growth of 'moth-eye' ZnO nanostructures on Si (111), c-Al_2O_3, ZnO and steel substrates by pulsed laser deposition", *Physica Status Solidi C*, Vol. 10, pp. 1317–1321.Sarhaddi, F., Farahat, S., Ajam, H., Behzadmehr, A., and Mahdavi Adeli, M., (2010), "An improved thermal and electrical model for a solar photovoltaic thermal (PV/T) air collector", *Applied Energy*, Vol. 87, pp. 2328–2339.

Satsangia Vibha, R., Saroj, K., Singha Aadesh, P., Rohit, S., and Sahab, D., (2008), "Nanostructured hematite for photoelectrochemical generation of hydrogen", *International Journal of Hydrogen Energy*, Vol. 33, pp. 312–318.

Schlesinger, M., and Paunovic, M., (2000), *Modern Electroplating*, John Wiley & Sons, Inc., New York.

Schnitzer, H., Christoph, B., and Gwehenberger, G., (2007), "Minimizing greenhouse gas emission through the application of solar thermal energy in industrial processes. Approaching zero emissions", *Journal of Cleaner Production*, Vol. 15, pp. 1271–1286.

Schön, J.H., Binder, G., and Bucher, E., (1994), "Performance and stability of some new high-temperature selective absorber systems based on metal/dielectric multilayers", *Solar Energy Materials and Solar Cells*, Vol. 33, pp. 403–416.

Sebastian, P.J., Quintana, J., and Avila, F., (1997), "Retention of the high optical absorptance in thermally aged black chrome on variably sensitized Cu", *Solar Energy Materials and Solar Cells*, Vol. 45, pp. 65–74.

Selvakumar, N., and Barshilia, H.C., (2012), "Review of physical vapor deposition (PVD) spectrally selective coatings for mid- and high-temperature solar thermal applications", *Solar Energy Materials and Solar Cells*, Vol. 98, pp. 1–23.

Selvakumar, N., Barshilia, H.C., Rajam, K.S., and Biswas, A., (2010), "Structure, optical properties and thermal stability of pulsed sputter deposited high temperature HfO_x/Mo/HfO_2 solar selective absorbers", *Solar Energy Materials and Solar Cells*, Vol. 94, pp. 1412–1420.

Selvakumar, N., Manikandanath, N.T., Biswas, A., and Barshilia, H.C., (2012), "Design and fabrication of highly thermally stable HfMoN/HfON/Al_2O_3 tandem absorber for solar thermal power generation applications", *Solar Energy Materials and Solar Cells*, Vol. 102, pp. 86–92.

Selvakumar, N., Santhoshkumar, S., Basu, S., Biswas, A., and Barshilia, H.C., (2013), "Spectrally selective CrMoN/CrON tandem absorber for mid-temperature solar thermal applications", *Solar Energy Materials and Solar Cells*, Vol. 109, pp. 97–103.

Seo, W.S., Lee, J.H., Sun, X., Suzuki, Y., Mann, D., Liu, Z., Terashima, M., Yang, P.C., McConnell, M.V., Nishimura, D.G., and Dai, H., (2006), "FeCo/graphitic-shell nanocrystals as advanced magnetic-resonance-imaging and near-infrared agents", *Nature Materials*, Vol. 5, pp. 971–976.

Seraphin, B.O., (1979), *Solar Energy Conversion: Solid-State Physics Aspects: Topics in applied Physics*, Springer, Berlin, Germany.

Shackelford, J.F., and Alexander, W., (2001), *Materials Science and Engineering Handbook*, 3rd Edition, CRC Press LLC.

Shah, S.I., (1995), *Handbook of Thin Film Process Technology*, IOP Publishing Ltd.

Shan, Z., Liu, Y., Chen, Z., Warrender, G., and Tian, J., (2008), "Amorphous Ni – S – Mn alloy as hydrogen evolution reaction cathode in alkaline medium", *International Journal of Hydrogen Energy*, Vol. 33, pp. 28–33.

Sharma, A., Varshney, M., Shin, H.-J., Kumar, Y., Gautam, S., and Chae, K.H., (2014), "Monoclinic to tetragonal phase transition in ZrO_2 thin films under swift heavy ion irradiation: Structural and electronic structure study", *Chemical Physics Letters*, Vol. 592, pp. 85–89.

Shinn, M., Hultman, L., and Barnett, S.A., (1992), "Growth, structure, and microhardness of epitaxial TiN/NbN superlattices", *Journal of Materials Research*, Vol. 7, pp. 901–911.

Shumway, A., Jaworske, A., and Jaworske, D.A., (2003), "Solar selective coatings for high temperature applications", *AIP Conference Proceedings*, Vol. 654, pp. 65–70.

Siegfried, D.E., Cook, D., and Glocker, D., (1996), "Reactive Cylindrical Magnetron Deposition of Titanium Nitride and Zirconium Nitride Films", 39th Annual Technical Conference Proceeding of the Society of Vacuum Coater, pp. 97–101.

Siu, G.G., Stokes, M.J., and Liu, Y., (1999), "Variation of fundamental and higher-order Raman spectra of ZrO_2 nanograins with annealing temperature", *Physical Review B*, Vol. 59, pp. 3173–3179.

Skoog, D.A., West, D.M., Holler, F.J., and Crouch, S.R., (2004), *Fundamnetal of Analytical Chemistry*, 8th Edition, Cengage Learning India Private Limited.

Solanki, C.S., (2013), *Solar Photovoltaics Fundamentals, Technologies and Applications*, 2nd Edition, PHI Learning Private Limited, New Delhi, India.

Song, Y., and Chin, D.-T., (2002), "Current efficiency and polarization behavior of trivalent chromium electrodeposition process", *Electrochimica Acta*, Vol. 48, pp. 349–356.

Spitz, J., Van Danh, T., and Aubert, A., (1979), "Chromium black coatings for photothermal conversion of solar energy, part I: Preparation and structural characterization", *Solar Energy Materials*, Vol. 1, pp. 189–200.

Stern, M., and Geary, A.L., (1957), "Electrochemical polarization", *Journal of the Electrochemical Society*, Vol. 104, p. 559.

Sukhatme, S.P., and Nayak, J.K., (2008), *Solar Energy Principle of Thermal Collection and Storage*, 3rd Edition, Tata McGraw Hill Education Private Limited, New Delhi, India.

Surviliene, S., Orlovskaja, L., and Biallozor, S., (1999), "Black chromium electrodeposition on electrodes modified with formic acid and the corrosion resistance of the coating", *Surface and Coatings Technology*, Vol. 122, pp. 235–241.

Suzer, S., Kadirgan, F., Sohmen, H.M., Wetherilt, A.J., and Ture, I.E., (1998), "Spectroscopic characterization of Al_2O_3-Ni selective absorbers for solar collectors", *Solar Energy Materials and Solar Cells*, Vol. 52, pp. 55–60.

Sweet, J.N., Pettit, R.B., and Chamberlain, M.B., (1984), "Optical modeling and aging characteristics of thermally stable black chrome solar selective coatings", *Solar Energy Materials*, Vol. 10, pp. 251–286.

Sylwestrowicz, W., and Hall, E.O., (1951), "The deformation and ageing of mild steel", *Proceedings of the Physical Society London, Section B*, Vol. 64, p. 495.

Tabor, H., (1956), "Selective radiation: I. Wavelength discrimination, II. Wavefront discrimination", *Bulletin of the Research Council of Israel*, Vol. 5A, pp. 119–134.

Tabor, H., (1961), "Solar collectors, selective surfaces and heat engines", *Proceedings of the National Academy of Science of the United States of America*, Vol. 47, pp. 1271–1278.

Tam, K.H., Cheung, C.K., Leung, Y.H., Djurisic, A.B., Ling, C.C., Beling, C.D., *et al.*, (2006), "Defects in ZnO nanorods prepared by a hydrothermal method", *The Journal of Physical Chemistry B*, Vol. 110, pp. 20865–20871.

Tang, W.T., Ying, Z.F., Hu, Z.G., Li, W.W., Sun, J., Xu, N., and Wu, J.D., (2010), "Synthesis and characterization of HfO_2 and ZrO_2 thin films deposited by plasma assisted reactive pulsed laser deposition at low temperature", *Thin Solid Films*, Vol. 518, pp. 5442–5446.

Teixeira, V., Sousa, E., Costa, M.F., Nunes, C., Rosa, L., Carvalho, M.J., Collares-Pereira, M., Roman, E., and Gago, J., (2001), "Spectrally selective composite coatings of Cr-Cr_2O_3 and Mo-Al_2O_3 for solar energy applications", *Thin Solid Films*, Vol. 392, pp. 320–326.

Tharamani, C.N., and Mayanna, S.M., (2007), "Low-cost black Cu-Ni alloy coatings for solar selective applications", *Solar Energy Materials and Solar Cells*, Vol. 91, pp. 664–669.

Thirugnanasambandam, M., Iniyan, S., and Goic, R., (2010), "A review of solar thermal technologies", *Renewable and Sustainable Energy Reviews*, Vol. 14, pp. 312–322.

Trasatti, S., (1972), "Work function, electronegativity, and electrochemical behavior of metals: III. Electrolytic hydrogen evolution in acid solutions", *Journal of Electroanalytical Chemistry*, Vol. 39, pp. 163–184.

Trasatti, S., (1976), *Advances in Electrochemistry and Electrochemical Engineering*, Vol. 10, Gerisher, H., and Tobias, C.W., (Eds.), John Wiley and Sons, New York, p. 298.

Turgut, Z., Scott, J.H., Huang, M.Q., Majetich, S.A., and McHenry, M.E., (1998), "Magnetic properties and ordering in C-coated Fe_xCo_{1-x} alloy nanocrystals", *Journal of Applied Physics*, Vol. 83, p. 6468.

Usmani, B., and Dixit, A., (2016a), "Spectrally selective response of ZrO_x/ZrC-ZrN/Zr absorber-reflector tandem structures on stainless steel and copper substrates for high temperature solar thermal applications", *Solar Energy*, Vol. 134, pp. 353–365.

Usmani, B., and Dixit, A., (2016b), "Impact of corrosion on microstructure and mechanical properties of ZrO_x/ZrC-ZrN/Zr absorber – reflector tandem solar selective structures", *Solar Energy Materials and Solar Cells*, Vol. 157, pp. 733–741.

Usmani, B., and Harinipriya, S., (2013), "Characterization black chrome films in presence and absence of graphite encapsultaed FeCo nanoparticles by electrodepoaition technique for solar thermal applications, *ECS Transactions*, Vol. 53, pp. 47–61.

Usmani, B., Vijay, V., Chhibber, R., Chandra, L., and Dixit, A., (2016a), "Zirconium carbide-nitride composite matrix based solar absorber structures on glass and aluminum substrates for solar thermal applications", *ISES Proceeding* (in press).

Usmani, B., Vijay, V., Chhibber, R., and Dixit, A., (2016b), "Optimization of sputtered zirconium films infrared reflector in spectrally selective solar absorbers", *Journal of Thin Solid Films* (in revision).

Usmani, B., Vijay, V., Chhibber, R., and Dixit, A., (2016d), "Solar performance analysis of ZrO_x/ZrC-ZrN/Zr absorber-reflector tandem structures under extreme thermal environment", *Springer Proceeding* (under review).

Valayapetre, M., Inal, O.T., Murr, L.E., Torma, A.E., and Rosenthal, A., (1979), "Microstructural and mechanical property evaluation of black-chrome coated solar collectors", *Solar Energy Materials*, Vol. 2, pp. 177–199.

Valleti, K., Murali Krishna, D., and Joshi, S.V., (2014), "Functional multi-layer nitride coatings for high temperature solar selective applications", *Solar Energy Materials and Solar Cells*, Vol. 121, pp. 14–21.

Van De Walle, C.G., (2000), "Hydrogen as a cause of doping in zinc oxide", *Physical Review Letters*, Vol. 85, pp. 1012–1015.

Van De Walle, C.G., (2001), "Defect analysis and engineering in ZnO", *Physica B*, Vol. 308–310, pp. 899–903.

Van de Walle, C.G., (2002), "Strategies for controlling the conductivity of wide-band-gap semiconductors", *Physica Status Solidi (b)*, Vol. 229, pp. 221–228.

Vaz, F., Cerqueira, P., Rebouta, L., Nascimento, S.M., Alves, E., Goudeau, P., *et al.*, (2004), "Structural, optical and mechanical properties of coloured TiN_xO_y thin films", *Thin Solid Films*, Vol. 447–448, pp. 449–454.

Velumani, S., Mathew, X., Sebastian, P.J., Narayandass, S.K., and Mangalaraj, D., (2003), "Structural and optical properties of hot wall deposited CdSe thin films", *Solar Energy Materials and Solar Cells*, Vol. 76, pp. 347–358.

Veprek, S., and Veprek-Heijman, M., (2012), "Limits to the preparation of superhard nanocomposites: Impurities, deposition and annealing temperature", *EMRS 2011 Symposium Quarterly, Thin Solid Films*, Vol. 522, pp. 274–282.

Veziroglu, T.N., and Barbir, F., (1995), "Transportation fuel-hydrogen", *Energy Technology and the Environment*, Vol. 4. pp. 2712–2730.

Vien, T.K., Sella, C., Lafait, J., and Berthier, S., (1985), "Pt-Al_2O_3 selective cermet coatings on superalloy substrates for photothermal conversion up to 600°C", *Thin Solid Films*, Vol. 126, pp. 17–22.

Vinogradov, S.N., and Linnel, R.H., (1971), *Hydrogen Bonding*, Chaps. 2, 3, 4, and 6, Van Nostrand Reinhold, New York.

Wäckelgård, E., (1998), "Characterization of black nickel solar absorber coatings electroplated in a nickel chlorine aqueous solution", *SEM*, Vol. 56, p. 35.

Wäckelgård, E., Mattsson, A., Bartali, R., Gerosa, R., Gottardi, G., Gustavsson, F., Laidani, N., Micheli, V., Primetzhofer, D., and Rivolta, B., (2015), "Development of W – SiO_2 and Nb – TiO_2 solar absorber coatings for combined heat and power systems at intermediate operation temperatures", *Solar Energy Materials and Solar Cells*, Vol. 133, pp. 180–193.

Waite, M.M., Shah, S.I., and Glocker, D.A., (2010), "Sputtering sources", *SVC Bulletin Spring*, pp. 42–50.

Waits, R.K., (1978), "Planar magnetron sputtering", *Journal of Vacuum Science and Technology*, Vol. 15, pp. 179–187.

Wang, J., (2006), *Analytical Electrochemistry*, 3rd Edition, John Wiley & Sons, Inc., Hoboken, NJ.

Wang, Q., Yao, Y., Shen, Z., Hu, M., and Yang, H., (2023), "Concentrated solar power tower systems coupled locally with spectrally selective coatings for enhancement of solar-thermal conversion and economic performance", *Green Energy and Resources*, Vol. 1.

Wang, X., Lee, E., Xu, C., and Liu, J., (2021), "High-efficiency, air-stable manganese–iron oxide nanoparticle-pigmented solar selective absorber coatings toward concentrating solar power systems operating at 750 °C", *Journal of Materials Today Energy*, Vol. 19, p. 100609.

Wang, Y., Lu, Z.W., Gao, X.P., Hu, W.K., Jiang, X.Y., Qu, J.Q., *et al.*, (2005), "Electrochemical properties of the ball-milled $LaMg_{10}Ni_{2-x}Al_x$ alloys with Ni powders (x = 0, 0.5, 1 and 1.5)", *Journal of Alloys and Compounds*, Vol. 389, pp. 290–295.

Wiesendanger, R., (1994), *Scanning Probe Microscopy and Spectroscopy: Methods and Applications*, Cambridge University Press.

Williamson, G.B., and Smallman, R.C., (1956), "Dislocation densities in some annealed and cold-worked metals from measurements on the X-ray debye scherrer spectrum", *Philosophical Magazine*, Vol. 1, p. 34.

Window, B., Ritchie, I.T., and Cathro, K., (1979), "A study of selective electroplated chrome blacks", *Thin Solid Films*, Vol. 57, pp. 309–314.

Window, B., and Savvides, N., (1986), "Charged particle fluxes from planar magnetron sputtering sources", *Journal of Vacuum Science & Technology A: Vacuum, Surfaces, and Films*, Vol. 4, p. 196.

Wolcott, A., Smith, W.A., Kuykendall, T.R., Zhao, Y., and Zhang, J.Z., (2009), "Photoelectrochemical study of nanostructured ZnO thin films for hydrogen generation from water splitting", *Advanced Functional Materials*, Vol. 19, pp. 1849–1856.

Wu, J.-J., and Liu, S.-C., (2002), "Catalyst-free growth and characterization of ZnO nanorods", *The Journal of Physical Chemistry B*, Vol. 106, pp. 9546–9551.

Wu, L., Gao, J., Liu, Z., Liang, L., Xia, F., and Cao, H., (2013), "Thermal aging characteristics of CrN_xO_y solar selective absorber coating for flat plate solar thermal collector applications", *Solar Energy Materials and Solar Cells*, Vol. 114, pp. 186–191.

Wu, S., Cheng, C.H., Hsiao, Y.J., Juang, R.C., and Wen, W.F., (2016), "Fe_2O_3 films on stainless steel for solar absorbers", *Renewable and Sustainable Energy Reviews*, Vol. 58, pp. 574–580.

Wu, Y., Zheng, W., Lin, L., Qu, Y., and Lai, F., (2013), "Colored solar selective absorbing coatings with metal Ti and dielectric AlN multilayer structure", *Solar Energy Materials and Solar Cells*, Vol. 115, pp. 145–150.

Xiao, X., Miao, L., Xu, G., Lu, L., Su, Z., Wang, N., and Tanemura, S., (2011), "A facile process to prepare copper oxide thin films as solar selective absorbers", *Applied Surface Science*, Vol. 257, pp. 10729–10736.

Xinkang, D., Cong, W., Tianmin, W., Long, Z., Buliang, C., and Ning, R., (2008), "Microstructure and spectral selectivity of Mo – Al_2O_3 solar selective absorbing coatings after annealing", *Thin Solid Films*, Vol. 516, pp. 3971–3977.

Xu, C., Wang, X., and Liu, J., (2022, Spinel Cu-Mn-Cr oxide nanoparticle-pigmented solar selective coatings maintaining >94% efficiency at 750°C", *ACS Applied Materials & Interfaces*, Vol. 14, pp. 33211–33218.

Xu, M.H., Zhong, W., Wang, Z.H., Au, C., and Du, Y.W., (2013), "Highly stable FeCo/carbon composites: Magnetic properties and microwave response", *Physica E: Low-dimensional Systems and Nanostructures*, Vol. 52, pp. 14–20.

Yan, Y., Ahn, K.S., Shet, S., Deutsch, T., Huda, M., Wei, S.H., Turner, J., and Al-Jassim, M.M., (2007), "Band gap reduction of ZnO for photoelectrochemical splitting of water", *Proc. of SPIE*, Vol. 6650, p. 66500H-1.

Yan, Y., and Malen, J.A., (2013), "Periodic heating amplifies the efficiency of thermoelectric energy conversion", *Energy & Environmental Science*, Vol. 6, pp. 1267–1273.

Yin, Y., McKenzie, D.R., and McFall, W.D., (1996), "Cathodic arc deposition of solar thermal selective surfaces", *Solar Energy Materials and Solar Cells*, Vol. 44, pp. 69–78.

Yoo, Y.H., Le, D.P., Kim, J.G., Kim, S.K., and Van Vinh, P., (2008), "Corrosion behavior of TiN, TiAlN, TiAlSiN thin films deposited on tool steel in the 3.5 wt.% NaCl solution", *Thin Solid Films*, Vol. 516, pp. 3544–3548.

Yue, S., Yueyan, S., and Fengchun, W., (2003), "High-temperature optical properties and stability of Al_xO_y-AlN_x-Al solar selective absorbing surface prepared by DC magnetron reactive sputtering", *Solar Energy Materials and Solar Cells*, Vol. 77, pp. 393–403.

Zaban, A., Aruna, S.T., Tirosh, S., Gregg, B.A., and Mastai, Y., (2000), "The effect of the preparation condition of TiO_2 colloids on their surface structures", *The Journal of Physical Chemistry B*, Vol. 104, p. 4130.

Zajac, G., and Ignatiev, A., (1979), "High temperature optical and structural degradation of black chrome coatings", *Solar Energy Materials*, Vol. 2, pp. 239–247.

Zajac, G., Smith, G.B., and Ignatiev, A., (1980), "Refinement of solar absorbing black chrome microstructure and its relationship to optical degradation mechanisms", *Journal of Applied Physics*, Vol. 51, pp. 5544–5554.

Zein El Abedin, S., Borissenko, N., and Endres, F., (2004), "Electrodeposition of nanoscale silicon in a room temperature ionic liquid", *Electrochemistry Communications*, Vol. 6, pp. 510–514.

Zhang, Q.-C., (1998), "Stainless-steel – AlN cermet selective surfaces deposited by direct current magnetron sputtering technology", *Solar Energy Materials and Solar Cells*, Vol. 52, pp. 95–106.

Zhang, Q.-C., (1999), "High efficiency Al-N cermet solar coatings with double layer film structures", *Journal of Physics D-Applied Physics*, Vol. 32, pp. 1938–1944.

Zhang, Q.-C., Hadavi, M.S., Lee, K., and Shen, Y.G., (2003), "Zr – ZrO_2 cermet solar coatings designed by modelling calculations and deposited by dc magnetron sputtering", *Journal of Physics D: Applied Physics*, Vol. 36, pp. 723–729.

Zhang, Q.-C., and Mills, D.R., (1992a), "High solar performance selective surface using bi-sublayer cermet film structures", *Solar Energy Materials and Solar Cells*, Vol. 27, pp. 273–290.

Zhang, Q.-C., and Mills, D.R., (1992b), "New cermet film structures with much improved selectivity for solar thermal applications", *Applied Physics Letters*, Vol. 60, p. 545.

Zhang, Q.-C., Yin, Y., and Mills, D.R., (1996), "High efficiency Mo-Al_20_3 cermet selective surfaces for high-temperature application", *Solar Energy Materials and Solar Cells*, Vol. 40, pp. 43–53.

Zhang, Q.-C., Zhao, K., Zhang, B.-C., Wang, L.-F., Shen, Z.-L., Lu, D.-Q., Xie, D.-L., Zhou, Z.-J., and Li, B.-F., (1998a), "A cylindrical magnetron sputtering system for deposition metal-aluminum nitride cermet solar coatings onto batches of tubes", *Journal of Vacuum Science and Technology*, Vol. 16, pp. 628–632.

Zhang, Q.-C., Zhao, K., Zhang, B.C., Wang, L.F., Shen, Z.L., Zhou, Z.J., Lu, D.-Q., Xie, D.-L., and Li, B.F., (1998b), "New cermet solar coatings for solar thermal electricity applications", *Solar Energy*, Vol. 64, pp. 109–114.

Zhang, S., Li, L., and Kumar, A., (2009), *Materials Characterization Techniques*, CRC Press, Taylor & Francis Group, Boca Raton, London, New York.

Zhao, S., and Wäckelgård, E., (2006), "The optical properties of sputtered composite of Al-AlN", *Solar Energy Materials and Solar Cells*, Vol. 90, pp. 1861–1874.

Zhao, Z.W., Lei, W., Zhang, X.B., Wang, B.P., and Tay, B.K., (2009), "Nanocrystalline zirconium oxide thin films grown under low pulsed dc voltages", *Journal of Physics D: Applied Physics*, Vol. 42, p. 215408.

Zheng, L., Gao, F., Zhao, S., Zhou, F., Nshimiyimana, J.P., and Diao, X., (2013), "Optical design and co-sputtering preparation of high performance Mo – SiO_2 cermet solar selective absorbing coating", *Applied Surface Science*, Vol. 280, pp. 240–246.

Zhou, W.-X., Shen, Y., Hu, E.-T., Zhao, Y., Sheng, M.-Y., Zheng, Y.-X., S.-Y. Wang, Y.-P., Lee, C.-Z., Wang, D.W., Lynch, and Chen, L.-Y., (2012), "Nano-Cr-film-based solar selective absorber with high photo-thermal conversion efficiency and good thermal stability", *Optics Express*, Vol. 20, pp. 28953–28962.

Index

For Product Safety Concerns and Information please contact our EU representative GPSR@taylorandfrancis.com
Taylor & Francis Verlag GmbH, Kaufingerstraße 24, 80331 München, Germany

www.ingramcontent.com/pod-product-compliance
Lightning Source LLC
LaVergne TN
LVHW010605110826
845149LV00003B/774

* 9 7 8 1 0 3 2 9 1 5 7 7 7 *